T0140862

Jonas Stienen

Real-Time Auralisation of Outdoor Sound Propagation

Logos Verlag Berlin GmbH

λογος

Aachener Beiträge zur Akustik

Editors:
Prof. Dr.-Ing. Janina Fels
Prof. Dr. rer. nat. Michael Vorländer
Institute for Hearing Technology and Acoustics
RWTH Aachen University
52056 Aachen
www.akustik.rwth-aachen.de

Bibliographic information published by the Deutsche Nationalbibliothek

The Deutsche Nationalbibliothek lists this publication in the Deutsche Nationalbibliografie; detailed bibliographic data are available in the Internet at http://dnb.d-nb.de.

D 82 (Diss. RWTH Aachen University, 2022)

Logos Verlag Berlin GmbH 2023

ISBN 978-3-8325-5629-7

ISSN 2512-6008

Vol. 39

Logos Verlag Berlin GmbH
Georg-Knorr-Str. 4, Geb. 10,
D-12681 Berlin

Tel.: +49 (0)30 / 42 85 10 90
Fax: +49 (0)30 / 42 85 10 92
http://www.logos-verlag.de

Abstract

Auralisation describes the process of generating and presenting audible sound using computer programs and audio hardware. Since the result is perceived naturally by the human's auditory system, a demonstration by means of auralisation is easily comprehensible and does neither require background knowledge nor expertise. Producing auralisation under real-time constraints increases the implementation demands significantly. Real-time auralisation is required in applications that respond to user interaction, for example, in interactive *Virtual Reality* (*VR*) environments. Dynamically moving sound sources and receivers evoke a change in the perceived sound, and the corresponding result must be provided as quickly as possible. A feeling of immersion can be created if the response time of the system does not exceed perceptual thresholds and thus enables a plausible scene presentation. To achieve this emotional state, real-time auralisation must comply with the expected physical behavior. Because auralisation has gained much attention in room acoustics, the established concepts and approaches are insufficient, if applied to outdoor scenarios. This is due to the fact that simulation principles based on *Geometrical Acoustics* (*GA*) can determine specular sound reflection with ease, but only rudimentarily incorporate the contribution of diffracted sound. Nonetheless, the transmission of sound from a source to a receiver in an urban setting contains a significant contribution of reflected and diffracted components. Furthermore, highly dynamic virtual environments are time-variant. Traditional realisations neglect this characteristic, treating an acoustic environment as temporarily static on a frame-by-frame basis. Adaption to dynamic events is implemented by update routines performing sequences of time-invariant simulations, which is contradictory to the highly dynamic nature of outdoor scenarios. This misconception is addressed in this dissertation and an alternative solution is suggested. The realisation of a real-time auralisation application for outdoor environments represents a promising addition to current noise assessment procedures. It also delivers the foundation for auditory modality in *VR* regarding outdoor scenarios, serving as an audio-video tool for scientific investigations of urban sound environments, accounting for perceptual aspects.

Over the past decade, a comprehensive real-time auralisation framework named *Virtual Acoustics* (*VA*)[1] has been developed under the principles of open science. The modular structure enables to combine different components implementing acoustic simulation algorithms, rendering approaches and reproduction units. As an integral part of a real-time outdoor auralisation module, highly efficient geometrical propagation simulation algorithms have been implemented. These algorithms effectively solve the problem of connecting a sound source with a sound receiver via a geometrical mesh representing the urban environment. Based on an intermediate list of geometrical propagation paths, a transformation routine applies acoustic models, such as diffraction attenuation, along the given paths. The resulting accumulated *Transfer Functions* (*TFs*) are compared with measurements, and the investigations were extended to evaluate the integration of the acoustic modelling into the auralisation engine. Furthermore, an auralisation module employing a network of *Digital Signal Processing* (*DSP*) units has been proposed, efficiently rendering dynamic outdoor sound propagation under real-time processing constraints. The module consequently considers the time-variant Doppler shift as well as typical, frequency-dependent sound propagation phenomena, like diffraction, for each individual geometrical propagation path. A new central unit called *Single-Input Multiple-Output Variable Delay Line* (*SIMO-VDL*) is introduced for this purpose. Motivated by the challenging demands of real-time processing, a novel directional clustering approach is presented that produces a binaural output for the receiver using *Head-Related Impulse Response* (*HRIR*) convolution and an *Inter-aural Time Difference* (*ITD*) model. It has the capacity to incorporate a separate binaural directional feature for individual incoming wavefronts (i.e., individual propagation paths) with a marginal computation footprint. The procedure determines a limited and fixed number of *Principle Directions* (*PDs*) to economise on the computationally heavy time-domain convolution of *HRIRs*. The parameterisation of *PDs* and the assignment of individual wavefronts is based on an angular distance measure (k-means clustering), and an *ITD* correction accounts for azimuthal deviations. The feasibility of the proposed real-time auralisation system and the suitability to outdoor sound propagation are demonstrated in an interactive application using *VR* technology.

[1] `http://www.virtualacoustics.org`

Kurzfassung

Auralisierung beschreibt den Prozess der Erzeugung und Präsentation hörbaren Schalls mittels Computerprogrammen und Audio-Hardware. Weil das Resultat durch den menschlichen Hörapparat auf natürlichem Wege wahrgenommen wird, ist eine Demonstration durch Auralisierung unmittelbar verständlich und benötigt weder Hintergrundwissen noch Erfahrung. Eine solche Auralisierung unter Echtzeitbedingungen durchzuführen erhöht die Anforderungen an die Implementierung beträchtlich. Echtzeit-Auralisierung wird in Anwendungen benötigt, die auf Benutzereingaben reagieren, zum Beispiel, in interaktiven Umgebungen der Virtuellen Realität (*VR*). Sowohl dynamische, bewegte Schallquellen als auch Empfängerbewegungen rufen eine Änderung des wahrgenommenen Schalls hervor und erfordern, dass eine entsprechende Anpassung so schnell wie möglich bereitgestellt wird. Das Gefühl der Immersion kann erzeugt werden, wenn die Reaktionszeit des Systems perzeptive Schwellen unterschreitet und die Präsentation plausibel erscheint. Um diesen emotionalen Status zu erreichen, muss Echtzeit-Auralisierung mit dem erwarteten, physikalischen Verhalten übereinstimmen. Weil Auralisierung vordergründig in der Raumakustik Aufmerksamkeit erfahren hat, sind die bestehenden Konzepte und Ansätze unzureichend, wenn sie für Außenlärmsituationen eingesetzt werden. Der Grund besteht darin, dass die Prinzipien der Simulation, die auf der Geometrischen Akustik (*GA*) basieren, zwar Schallreflexionen mit Leichtigkeit bestimmen können, den Beitrag von gebeugtem Schall jedoch nur rudimentär einbeziehen. Nichtsdestotrotz beinhaltet die Schallübertragung von einer Quelle zu einem Empfänger eines urbanen Schauplatzes einen signifikanten Anteil von reflektiertem und gebeugtem Schall. Darüber hinaus sind hochdynamische virtuelle Umgebungen *zeitvariant*. Traditionelle Realisierungen vernachlässigen dieses Merkmal und betrachten eine akustische Umgebung auf der Basis von Einzelbildfolgen als temporär statisch. Dabei ist die Adaption dynamischer Ereignisse als Aktualisierungsroutine implementiert, die Sequenzen zeit*in*varianter Simulationen durchführt, welche im Widerspruch zu der hohen Dynamik von Außenlärmsituationen steht. Diese falsche Annahme wird in der vorliegenden Dissertation diskutiert und eine alternative Lösung wird vorgeschlagen. Die Umsetzung einer Echtzeit-Auralisierungsanwendung für urbane Umgebungen stellt eine vielversprechende Erweiterung für die gegenwärtigen Verfahren der Lärmbeurteilung dar. Zusätzlich liefert ein audio-visuelles Werkzeug die Grundlagen für die akustische Modalität in der *VR* bezüglich urbaner Szenarien, welche als wertvoll für wissenschaftliche Untersuchungen urbaner Schallumgebungen unter Berücksichtigung perzeptiver Aspekte erachtet wird.

Im Laufe des vergangenen Jahrzehnts wurde ein umfangreiches Software-Framework namens *Virtual Acoustics* (*VA*)[2] für die Echtzeit-Auralisierung unter den Prinzipien der offenen Wissenschaft ("open science") entwickelt. Die modulare Struktur erlaubt die Kombination verschiedener Komponenten, die akustische Simulationalgorithmen, Rendering-Ansätze und Reproduktionseinheiten umsetzen. Als integraler Bestandteil eines Echtzeit-Auralisierungsmoduls für Außenbereiche wurden hocheffiziente geometrische Wellenausbreitungsalgorithmen implementiert. Diese lösen wirksam das Problem, eine Schallquelle mit einem Schallempfänger über Flächen und Kanten eines geometrischen, polygonalen urbanen Umgebungsmodels zu verbinden. Basierend auf einem Zwischenresultat geometrischer Schallausbreitungspfade wird eine Transformationsroutine eingesetzt, welche verschiedene akustische Modelle, wie Beugungsdämpfung entlang des Pfades berücksichtigt. Die resultierenden, akkumulierten Transferfunktionen (*TFs*) wurden mit Messungen verglichen. Untersuchungen wurden angestellt, um die Evaluierung der Integration akustischer Modelle in die Auralisierungseinheit durchzuführen. Außerdem wurde ein Netzwerk digitaler Signalverarbeitungseinheiten (*DSP*) vorgeschlagen, welche die dynamische Schallausbreitung im Freien unter Gesichtspunkten der Echtzeitverarbeitung effizient abbildet. Konsequenterweise wird der zeitvariante Doppler-Effekt sowie typische, frequenzabhängige Schallausbreitungsphänomene, wie beispielsweise Beugung, für jeden geometrischen Ausbreitungspfad individuell berücksichtigt. Eine neue, zentrale Verarbeitungseinheit, die variable Verzögerungsleitung mit einem Eingang und beliebig vielen Ausgängen (*SIMO-VDL*) wird zu diesem Zweck eingeführt. Motiviert durch die herausfordernden Anforderungen an die Echtzeitverarbeitung wird ein neuartiger Ansatz zur Bündelung von richtungsabhängigem Schalleinfall präsentiert, der eine binaurale Ausgabe für einen Hörer erzeugt, welcher auf der Faltung von kopfbezogenen Impulsantworten (*HRIRs*) sowie einem interauralen Zeitversatzmodell (*ITD*) beruht. Das Verfahren besitzt die Kapazität separate binaurale Richtungseigenschaften für individuelle einfallende Schallwellen (z.B., von individuellen Schallausbreitungspfaden) mit einem marginalen Rechenaufwand einzubeziehen. Dabei wird zunächst eine beschränkte konstante Anzahl an Leitrichtungen (*PDs*) bestimmt, um Einsparungen der rechentechnisch

[2] `http://www.virtualacoustics.org`

aufwändigen Zeitbereichsfaltung der *HRIRs* zu erzielen. Die Parametrisierung der *PDs* und die Zuordnung individueller Wellenfronten basiert auf einer Winkeldistanzmetrik ("k-means clustering") und die Korrektur der *ITD* berücksichtigt die Abweichungen im Azimut. Die Machbarkeit des vorgeschlagenen Echtzeit-Auralisierungssystems und die Eignung für die Schallausbreitung im Freien werden in einer interaktiven Applikation demonstriert, die auf *VR*-Technologien basiert.

For the young who are inspiring me, the old who ground me
and my dearest who holds everything together.

Contents

Glossary

Acronyms

BEM	Boundary Element Method.
BIR	Binaural Impulse Response.
BRAS	Benchmark for Room Acoustic Simulations.
BT	Biot-Tolstoi.
BTM	Biot-Tolstoi-Medwin.
BTMS	Biot-Tolstoi-Medwin-Svensson.
CAVE	CAVE Automated Virtual Environment.
CPU	Central Processing Unit.
CTC	Cross-Talk Cancellation.
DC	Direct Current.
DFT	Discrete Fourier Transform.
DSP	Digital Signal Processing.
ESEA	Equation System of Equal Angles.
EU	European Union.
FD	Fractional Delay.
FDTD	Finite-Difference Time Domain.
FEM	Finite Element Method.
FIR	Finite Impulse Response.
FT	Fourier Transform.
GA	Geometrical Acoustics.

GIS	Geographic Information System.
GTD	Geometrical Theory of Diffraction.
HMD	Head-Mounted Display.
HOA	Higher-Order Ambisonics.
HRIR	Head-Related Impulse Response.
HRTF	Head-Related Transfer Function.
IEM	Image Edge Model.
IIR	Infinite Impulse Response.
IR	Impulse Response.
ISO	International Organization for Standardization.
ITD	Inter-aural Time Difference.
JSON	JavaScript Object Notation.
LTI	Linear Time-Invariant.
mESEA	modified Equation System of Equal Angles.
MIM	Mirror Image Model.
MULADD	Multiplication-And-Addition.
PC	Personal Computer.
PD	Principle Direction.
RIR	Room Impulse Response.
RMS	Root Mean Square.
SIMO-VDL	Single-Input Multiple-Output Variable Delay Line.
SNR	Signal-To-Noise Ratio.
TF	Transfer Function.
TOA	Time of Arrival.
UTD	Uniform Theory of Diffraction.
VA	Virtual Acoustics.
VBAP	Vector-Base Amplitude Panning.
VDL	Variable Delay Line.

VR	Virtual Reality.
WFS	Wave Field Synthesis.
WHO	World Health Organization.

1 Introduction

Noise is a matter that concerns everyone. It is interpreted as undesired sound perceived by a living being and hence has a negative connotation. Its connection to our well-being and the implications on health and mortality is statistically recorded and scientifically discussed, most prominently in the publications by the World Health Organization *(WHO)* [Wor99; Wor11]. Naturally, there is a broad interest to understand the mechanics behind noise, to gain knowledge about the influencing factors and to apply counter-actions [Kan06]. The context is manifold, and touches many disciplines. Architects and city planers, engineers, politicians, psychologists, physicists and physicians are all involved with noise issues in their own, specialised ways. However, one general problem can be identified: each discipline quantifies, interprets and handles acoustic matters in a distinct fashion. Dealing with standard acoustic metrics requires a certain level of understanding for a correct interpretation, which counteracts an intuitively comprehensible description. To give an example, it could be of interest to assess the effect of a new roadside noise barrier on the perceived traffic noise in a residential area. However, it is rather difficult to convince affected residents that a noise level mitigation of 3 dB delivers a substantial reduction and justifies a financial investment. In contrast, if the involved people or their representatives have the option to perceive the difference with their own ears, the implications are related to this specific experience. An interactive demonstration that presents the given context and provides the freedom to move – just like in reality – is one way to support the mutual understanding and can be achieved by means of responsive auralisation if certain aspects are considered.

This thesis focuses on a technical approach to make virtual interactive urban environments perceivable by the auditory senses. Such a *real-time auralisation system* directly aims at the misconception issues among laymen and interest groups, since it intends to transform a virtual scenario of a potentially real situation into a representation that can be interpreted intuitively. Auralisation does not require background knowledge in order to understand the demonstrated content, as it is perceivable in a natural fashion. An interactive system encourages

Figure 1.1: Real-time auralisation is used in the *aixCAVE*, the interactive *VR* system of the RWTH Aachen University.
Left image ©Virtual Reality Group, RWTH Aachen University.

to explore, investigate and even modify the environment with the benefit of an immediate response. It has the potential to provide assurance or identify problems in real-time.

The motivation to make progress in the research area of real-time auralisation in general and outdoor sound in particular is given by the lack of comprehensive and versatile tools to investigate the perceptual aspect of urban sound generation and propagation on a large scale. An approach is presented to close in on the mutual understanding of the expert party on the acoustic simulation side and the community, which ultimately has to live with the result. This work is dedicated to the invention and implementation of the required algorithms. In particular, propagation simulation approaches that deliver a sufficient result for the auralisation of urban environments is presented, which is computationally efficient and integrates reflections and diffractions. Furthermore, dynamic scenarios require dynamic rendering that provides a continuous auralisation output. Due to the fluctuation of propagation paths when simulating the acoustic transmission from sources to receivers in motion, an investigation of the current diffraction models is conducted in the context of real-time auralisation focusing on the provided sound field continuity. An approach is presented to avoid loudness steps in the rendering result by solving the problem of instantaneously appearing and disappearing direct and reflected sound field components in a dynamic scenario. Based on the described toolchain, the path for succeeding progress is cleared. The developed auralisation application, which is publicly available[3], forms the basis to initiate the establishment and evaluation of perceptually important aspects of complex dynamic urban environments. The framework can be employed in practical projects, either with a scientific background as *user studies*, or in a

[3] *Virtual Acoustics* (*VA*), `http://www.virtualacoustics.org`

Figure 1.2: Hyde park corner, London, England. Traffic simulation visualisation with 3D buildings and dynamically moving vehicles on the traffic lanes.

practical multi-physics applications providing acoustic feedback (cf. Figure 1.1 and Figure 1.2).

In particular, the role of interactive user studies must be outlined. The presentation of slightly varying auralisations offers great potential for comparison tests. Given the possibility to switch between different setups or virtual locations in the fraction of a second, an immediate state of the sound field can be investigated in terms of *perceptual differences and thresholds*. These kind of experiments are relevant to quantify noise control measures and deliver insight into the impact of mitigation aspects. Especially if the participants are left unaware of the background of the study, bias can be reduced and the quality of the findings are improved.

While such testing environments can be implemented using pre-rendered audio tracks, the provision of an interactive dimension in terms of free movement requires a working reactive real-time auralisation system. It is conceivable to limit interactivity to a certain degree to avoid the employment of a comprehensive auralisation application, for example, if rotational user movement is sufficient. The limitation of motion can be feasible to reproduce the acoustic quality of a scenario at a selection of fixed locations without the need of a real-time auralisation module. However, constraining the freedom to move in the context of outdoor scenarios imposes an unnatural restriction that deteriorates the immersive capability of the presentation. Furthermore, human perception relies on all senses, including vision, which is particularly important in an urban environment. Consequently, it is deemed appropriate to present virtual urban scenarios by an audio-video Virtual Reality *(VR)* system for a plausible user experience, which incorporates user movements beyond head rotation. This fact underlines the requirement of a real-time auralisation application supporting the auditory sense in every aspect, especially regarding user studies and informal demonstrations (see, e.g., [LVK02]).

In order to make a virtual situation audible by means of auralisation, it is imperative to comply with physical principles of sound generation, sound propagation and the mechanics of human perception. Models and algorithms must be carefully evaluated and simplifications require justification. Real-time capability can only be provided by innovative system design and sophisticated acceleration methods, which potentially introduce perceivable differences if compared to a reference. However, to find an appropriate reference in this context is a complicated task, since many aspects are influencing the result [BJ97; Jek04]. Responding to the inner reference, for example, addresses the user's expectation based on individual experience. In contrast, a reference that incorporates either real-space recordings, a computer-generated auralisation including measured components (e.g., a Room Impulse Response *(RIR)*) or a high-resolution simulation of a virtual environment is advantageous to steer the focus of an experiment. On the one hand, the reduction of complexity in a user study decreases side effects that negatively influence the results. On the other hand, a greater distance to reality deteriorates the ecological validity of the experiment. Because extensive user studies are costly, the achievement of a large sample sizes is challenging and makes classical statistical analysis vulnerable, as the methods only gain accuracy with increasing sample number. Consequently, the quality of user studies profits, if conducted in a VR system. Here, the human's attention is not per-se restricted to a certain objective. The participant can be assigned an activity, for example, a daily routine task solving an arithmetical problem or paying attention to the contents of a speech [MVS19; Ehr+20]. As a result, experiments can be designed to guide the participants and keep them occupied yet indirectly measuring the influence of an acoustic intervention. This can be achieved, for example, by a questionnaire or interview after the experiment, or by observing the participant's performance. This rationalises the practical value of VR systems and the demand to provide real-time auralisation.

On a technical note, if real-time auralisation is considered in the context of user studies and demonstrations, limitations must be derived from the specifications of the desired system. For example, it is usually predetermined if headphones

Figure 1.3: Alexanderplatz, Berlin, Germany. Virtual model based in CityGML with 3D buildings enriched by valuable meta information with regards to auralisation, like traffic lanes.

or loudspeakers are required. This decision guides the employment of methods for the technical implementation of audio rendering and reproduction. Furthermore, the degree of interactivity greatly influences the configuration of modules that must be updated according to the user input and guides the specification of the real-time constraints for certain components of the system. For example, requirements are largely reduced, if the user is only allowed to rotate the head. In this case, even pre-rendered multi-channel tracks can be engaged by choosing audio formats that maintain the rotational freedom for the listener. If the user is restricted to movements within a narrowed area or along a predefined trajectory pre-calculations present a valid option. In contrast, if arbitrary motion (rotation *and* translation) in a vast virtual environment is desired, pre-calculation is infeasible as the number of states that must be covered becomes unmanageable.

Real-time auralisation overcomes this problem by calculating only what is currently necessary at the cost of a challenging technical realisation.

The overarching objective of the efforts put into real-time auralisation of outdoor sound propagation is to a) provide new possibilities of noise assessment and b) achieve an interactive virtual acoustic representation of a space that reflects reality in all its facets (i.e., outdoor acoustics for VR). An illustrative example is a virtual soundwalk in an arbitrarily chosen place based on real-world data, which complies with the expected soundscape at the given daytime regarding the volume of traffic, for instance, at the Alexanderplatz in Berlin, as visualised in Figure 1.3 using open geographic data that also includes traffic lanes and further metadata.

This thesis is concerned with the auralisation of outdoor sound under real-time conditions. Special attention is drawn to sound propagation simulation and the implication of dynamic aspects of virtual urban environments. A brief outline reads as follows:

The next chapter describes fundamentals of outdoor sound modelling by introducing the standard propagation models with an emphasis on relevant state-of-the-art diffraction models. Current assessment methods to quantify and predict outdoor noise are reviewed and an overview on simulation and auralisation described in the literature is provided.

In the third chapter, *sound propagation simulation* and its applicability for real-time auralisation is described. Special attention is drawn to Geometrical Acoustics *(GA)* methods and a particular approach is described in detail that is able to rapidly determine sound propagation paths in three-dimensional space featuring arbitrary reflections and diffractions. The approach has been realised and improved during the implementation of a comprehensive urban sound auralisation framework, as it is deemed particularly relevant for outdoor scenarios.

The fourth chapter presents a real-time audio rendering system based on a dedicated time-variant Digital Signal Processing *(DSP)* network. The proposed layout is specifically designed to handle many fast-moving sound sources and several receivers that can move freely, as anticipated in an acoustic outdoor scenario.

A fifth chapter deals with the feasibility of individual building blocks of the proposed concept, classifies the scope of the described auralisation system and verifies real-time operability by an example application. It also points out theoretical and practical limitations of the approach.

The final chapters conclude the work, summarise the important contributions and give an outlook.

2 Fundamentals of outdoor sound modeling

From a physical viewpoint, sound and vibration behaves similar everywhere. However, indoor and outdoor sound is interpreted fundamentally different. Outdoor sound consists of environmental and technical sound sources that cover the entire audible frequency range from 20Hz to 20kHz and span from vanishingly quiet to overwhelmingly high sound power levels. The characteristic open range causes sound to be spread in all directions, since acoustic energy is not generally redirected backwards by the majority of boundaries, as often found in confined spaces. In addition, formation of ground topology, composition and density of the built environment and large distances play an important role in the mitigation of noise levels.

While the propagation medium (air) usually does not change indoors, we expect temperature gradients and wind outside that significantly influence the perceived sound. An inhomogeneous propagation medium evokes sound refraction and thus wavefronts that no longer propagate in geometrically straight lines. Meteorological conditions, like wind and turbulence, introduce a time-variant component in the propagation medium. On a micro-time scale, wind results in a frequency shift depending on the relative direction with respect to a sound source *and* a receiver (Doppler). On a macro-time scale, conditions of temperature, relative humidity and static pressure can fluctuate significantly and are likely to impact sound propagation noticeably when put in direct comparison.

To realise a successful auralisation, it is important to bear the characteristics of outdoor sound in mind. Source-related properties, propagation phenomena and receiver characteristics must be modelled with sufficient accuracy to achieve the overall goal of creating a convincing virtual acoustic environment that complies with reality as close as possible [Vor11]. Assuming that sound behaves linear, it is permitted to superpose the individual contributions of each sound source to the sound field at a receiver location. This separation breaks down the sound modelling problem into a sum of individual source-receiver constellations, each represented by a function describing the point-to-point propagation component,

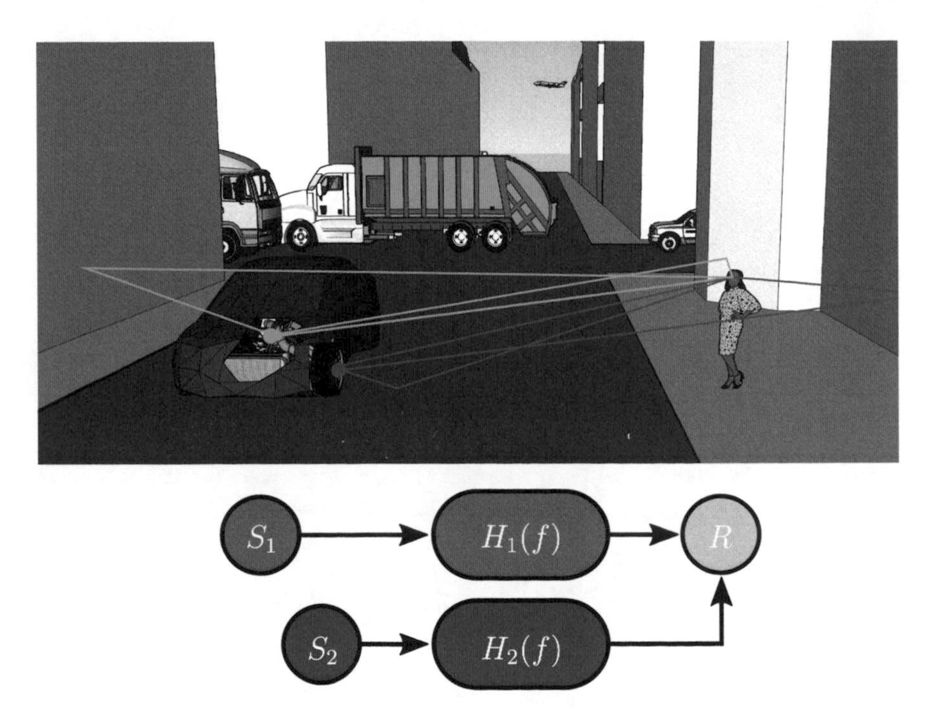

Figure 2.1: Virtual traffic noise scenario in an urban setting. Classical sound modelling discriminates various sound objects, as indicated for the schematic red car in the front. An engine sound source S_1 (green) and a tire sound source S_2 (purple) establish individual, geometrical propagation transfer paths with the receiver R via the environment. Sound propagation is regarded as a function of frequency, $H(f)$.

as depicted in Figure 2.1. The environment is usually regarded as *static* and sound transmission is based on the assumption of steady-state acoustic behaviour, at least for the corresponding time snapshot, in disregard of transient features. For a simplified mathematical handling of the various attenuation factors involved, the Fourier space is used and the propagation problem is considered in the frequency domain. Hence, it is described by a frequency-dependent transfer function $H(f)$ under linear, time-*in*variant conditions.

The steady-state sound propagation modelling has proven applicability in countless situations. However, regarding the auralisation of dynamic outdoor scenarios, the approach reaches its feasibility limits, as will be discussed in Section 2.6.

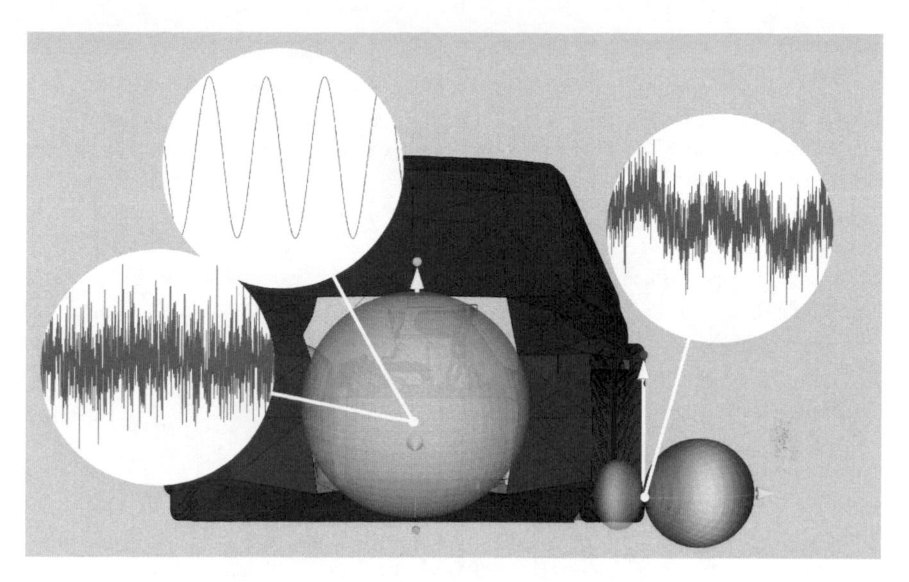

Figure 2.2: Example of an acoustic modelling of a virtual car for auralisation purposes. The object is separated into point sources of engine compartment and tires. The engine source is emitting a tonal and a noise signal, while the tire sources are only using a noise signal. The engine is assigned an omnidirectional directivity (valid particularly for lower frequencies). The tires have a more pronounced side lobe while dampening sound emitted to the contra-lateral side of the car.

2.1 Sound source characteristics

The characteristics of sound sources encountered in outdoor scenarios are diverse. Depending on the context, one can generally distinguish between human-made noise and environmental sounds [BKG11]. Examples for technical noise sources are traffic vehicles, railroad trains, aeroplanes, helicopters, drones, fans, pumps, industrial plants, wind turbines - to name a few. Other sources, that are linked to environmental sounds are, for example, wind hauling, leave rustling, thunder, as well as sounds from domestic or wildlife animals, like a dog's barking or a bird's chirping.

To construct a theoretical model of the variety of different sound sources, an acoustic excitation type and a radiation pattern must be defined. In general, sound sources are represented by infinitesimally small point sources, that radially emit a distinct (and unique) source signal [Vor11]. The assumption of a point source holds, if the real source's sound field behaves like a radial function that describes the emitted sound field sufficiently. For complex-shaped sound sources, a certain minimum distance to a receiver or any other obstacle from the source's location is required. The directional difference pattern is classified as the directivity of a sound source. Figure 2.2 shows an example of a virtual car that is represented by an engine compartment and tires, which are treated as separate sound sources. Each source has an individual, relative location and orientation, and is assigned a corresponding directivity. As an example, the engine signal is synthesised by mixing a tonal component and a noise component, while the tires are generating a noise signal, only.

The emitted signal that is propagated through the medium to a receiver is an important property of a sound source, because it carries acoustic information that generates a characteristic context of the soundscape. In contrast to an energetic noise assessment method, the auralisation approach *maintains the acoustic content* and delivers an audio feedback that enables the user to recognise the sounding object. This quality is both exciting and challenging, because the presented auralisation system can be significantly deteriorated in terms of authenticity and realism, if the provided content does not comply with the expectation. For example, if an accelerating or decelerating car does not deliver an acoustic signal that reflects a varying engine load, the scenario is perceived as unnatural.

2.2 Standard propagation models

Following the fundamental principle of outdoor sound modelling (cf. Figure 2.1), the transmission of acoustic waves from the source to the receiver requires a function $H(f)$ of frequency f [Vor11]. This function incorporates relevant acoustic phenomena and assumes a static scenario, at least for an instance in time. It describes the steady-state sound attenuation per frequency and is generally of sufficiently high resolution. To represent an energetic band spectrum, frequencies are grouped in the audible range at standardised centre values and referred to as octave-bands and third-octave bands, for example according to *ISO* 266 [ISO97]. The corresponding magnitudes are real-valued scalars that are normally scaled with a base-10 logarithmic function and referred to as decibels with the unit *dB*. In the context of sound propagation, the reference value can be chosen to present the induced modification making a comparison more comprehensible. Typically, a simplified scenario serves as a reference to remove influences of the source and the receiver, like a direct point-to-point transmission under free-field conditions. This way, the influences of source and receiver are equalised and the resulting spectrum is transferable.

The consideration of energetic band spectra is useful for noise assessment purposes, as will be discussed in Section 2.4. A more precise representation that incorporates coherent sound pressures requires a complex-valued spectrum that provides both the amplitude and the phase information of a given frequency. Such a spectrum contains the embedded temporal information and maintains the possibility to perform a transformation between time and frequency domain after Fourier (see, e.g., [OL07]). In particular, the superposition of similar sound waves reveals constructive and destructive interference depending on the frequency (or the wave length). This constellation produces comb filter transmission spectra that are apparent, for example, if a sound is reaching a receiver on a direct path and via a reflection. Because ultimately the human auditory system has the capacity to resolve coherent signals, a comprehensive auralisation approach must maintain this feature.

Apart from specular reflections, which add a quasi-identical signal to the direct sound, diffuse reflections contribute to the total sound field at the receiver as a result of scattered sound [Vor11]. Because the degree of diffuseness depends on the spatial dimension of surface structures, a pronounced frequency dependency can be expected where surfaces are rough with respect to the regarded wave length.

In the region where objects are large compared to the wave length, sound is subject to diffraction at discontinuities of surface boundaries. Especially in the absence of direct sound and specular reflections, diffracted sound is of great importance. This is the case, for example, if sound propagates over noise barriers or around buildings to a receiver located in the acoustic shadow region (sometimes also referred to as occlusion effect). However, the diffracted sound component is an equally important contributor in the illuminated region in order to maintain a continuous description of the overall sound field and thus it is a property that is of importance for real-time auralisation, as will be discussed in Section 3.4 and evaluated in Section 5.4.

Depending on the wave type that is emitted by a sound source, acoustic intensity decreases with distance equally for all frequencies. In theory, in the context of outdoor sound, point sources and line sources are relevant. However, considering the auralisation of dynamic environments, the great majority of sound sources can be modelled as a point source, which results in a spreading loss factor that is directly reciprocal to the distance and is also referred to as geometrical divergence or 1/r distance law.

Sound propagation outdoors is also subject to dampening by the propagation medium air, itself. The composition and condition has an absorbing effect on the transmitted acoustic signal over distance, excessively attenuating the high frequency content while leaving very low frequencies largely unaffected.

The Doppler shift is a time-variant acoustic phenomenon that is observed upon relative translational movements of the source *and/or* the receiver with respect to the medium. Towards the relative direction of movement, the emitted source signal into the medium is contracted, effectively shifting lower frequencies into a higher range. In the opposite direction, signals are expanded and the initial content is perceived as scaled towards lower frequencies. An acoustic sensor at rest receives the corresponding stretched or squeezed signals after the propagation time has elapsed. However, if the receiver is moving as well, another Doppler shift is caused, which affects the signal propagating in the medium. Handling the Doppler shift is not directly integrable into the propagation model of Figure 2.1, because it violates the assumption of a static environment and is in contradiction with the concept of a steady-state description. An approach to integrate Doppler shifts for an auralisation application stores the source-induced relative direction at a given point in time minus the elapsed propagation time, and evaluates the corresponding Doppler shift with respect to the current receiver location [SV18]. The frequency-dependent propagation attenuation must either be applied to a modified source signal according to the source-induced Doppler shift, or the source and receiver shifts can be combined making it necessary to scale the spectra of the propagation effects accordingly, as will be discussed in Section 4.7.2.

2.3 Acoustic diffraction models

The phenomenon of wave diffraction has been investigated scientifically for over a century now (Huygens, Fresnel and Kirchhoff, see [BW97]). As early as the 1900s, MacDonald formulated mathematical concepts to describe diffraction problems of electromagnetic waves [Mac02; Mac15]. Already in 1896, Sommerfeld discussed the mathematical question of diffraction around a thin wedge [Som96] regarding optical and acoustical application. The *Geometrical Theory of Diffraction* (*GTD*) evolved in 1957, which was described by Keller [KB51; Kel57; Kel62]. It has an certain proximity to the description of rays that refract close to edges of diffraction.

The publication of Biot and Tolstoi the same year can be interpreted as a milestone within the theory of diffraction and is referred to as the *Biot-Tolstoi* (*BT*) model [BT57]. It is the first publication to describe the solution of a diffraction problem in the time domain by transforming the wave equation into its angular modes in cylindrical coordinates. This interpretation aligns well with GA principles that interprets acoustic waves as rays or particles. The achievement paved the way for a series of comparisons and verifications regarding the different theories through measurement results [BM78; Bre78; KC80].

In 1981, Medwin expedited the Biot-Tolstoi *(BT)* method [Med81]. In his publication on the diffraction of sound in the acoustical shadow zone, a closed form of the BT formula is presented, referred to as *Biot-Tolstoi-Medwin* (*BTM*) model. The method predicts attenuation of sound barriers with arbitrary opening angles that are, in contrast to the original view, of finite extent. Other views on the matter evolved that followed different approaches, for example, a Huygens interpretation and multiple diffraction [MCJ82], as well as further comparative investigations based on simulations and measurements proving the weak points of the Kirchhoff approximation. These findings elevated the relevance of the BT diffraction formula and the extension by Medwin for fundamental investigations that require a high accuracy regarding acoustic impulse responses.

Beginning in the 1940's, the shadowing of sound sources through sound barriers was simultaneously examined by means of a pragmatic approach [Red40]. Especially important is *Maekawa*'s publication from 1968, which determines the diffraction of sound at an infinite half-plane as the reduction of sound energy by experimental determination in form of a function of detour and wavelength [Mae68]. The result is solely based on geometrical operations to identify the way around the boundary to the receiver, as discussed in more detail in Section 2.3.1. The approach was then extended and applied to all types of transmission paths, for example, around groups of houses or noise barriers. The amount of energy is subsequently accumulated, as specified by *ISO* 9613-2 [ISO96] (cf. also Section 2.5.1). To establish simplified procedures with reasonable approximation errors, modifications that combine theory and empiricism also rely on the approach of the specification of shadowing through sound barriers, such as that of Pierce in 1974 [Pie74].

While Maekawa's by now well-established model, usually referred to as rubber band model, has found application in noise immission simulation software (e.g., trough standards that make use of Maekawa's approximation, see Section 2.5), more accurate methods remained subject to scientific investigation. In 1999, Svensson et al. could significantly ameliorate the mathematical description of the effects of diffraction in both time and frequency domain [SFV99]. The Biot-Tolstoi-Medwin *(BTM)* diffraction method that has been improved by Medwin and was further extended by Svensson, who substitutes a diffraction edge by a model of secondary directional sources, is an integral time-domain solution to an acoustic diffraction problem of a finite-sized wedge. This feature naturally is very appealing, as polyhedral representations of virtual urban environments comprise finite elements. In contrast to the above-mentioned methods, the Biot-Tolstoi-Medwin-Svensson *(BTMS)* model is founded on an *acoustic* diffraction interpretation due to the physical derivation originating from the Huygens-Fresnel principle [MCJ82; SFV99]. The BTMS finds applications in the investigation of acoustic modelling technology, for example, comparison studies of fundamental problems [AP13; SSV11; Sch+12], scattering [SA12; Mar+18] and room acoustic modelling [SP09; TSK01; TK98]. The employment of BTMS for auralisation purposes is reported, for example, by Calamia et al. [CSF05; Cal09], Lokki et al. [LSS02] and Torres et al. [TRK04]. There is also literature describing successful integration into interactive systems while overcoming the shortcomings, for instance, the publications by Antani et al. [Ant+10; Ant+12]. Other contributions deal with the continuity problem at the transition areas of acoustical shadow and half-shadow zones [SC06]. BTMS diffraction also became an integral part of some interactive auralisation approaches with the help of various acceleration methods [AS10; CS07; CMS08].

A problematic property of the method that has been found troublesome in the context of dynamic outdoor auralisation, however, is the occurrence of singularities. It is mentioned that the BT formulation and the Medwin extension expose singularities, for example, at the values of initial arrival time [CK88; SFV99; TSK01] and in the vicinity of transitions zones [SC06; Cal09]. As has been pointed out earlier, continuity of the sound field is imperative for perception-appropriate dynamic auralisation. Additionally, reservations exist about the feasibility for virtual urban environments considering the increased computational demands. Due to these arguments, the detailed discussion of the complex mathematics of the BTMS has been put aside with reference to the available literature. The implementation of BTMS for the purpose of designing a real-time outdoor auralisation method is not considered in this work, without providing an evaluation comparing the existing auralisation methods with the concept proposed in this thesis (cf. Section 3.5).

In 2009, Taylor et al. used techniques of computer graphics to combine the *Uniform Theory of Diffraction* (*UTD*) formulation with the ray-frustum procedure, which expands more frusta at defined edges and identifies covered sources in the shadow zone [Tay+09a; Tay+09b]. It is stated that this approach is able to achieve sufficiently high update rates for real-time applications using regular, multicore processors, which is an appealing feature for interactive outdoor environments. The *UTD* blends well with the mirror image method and the combination of deterministic reflection and diffraction methods is highly relevant for outdoor auralisation problems [ESV21] (cf. Section 2.3.2). Further literature reporting on the integration of the *UTD* method in auralisation applications is presented, for example, by Schissler et al. [SM16b].

A fundamentally different yet noteworthy approach to diffraction modelling is presented by Stephenson and Svensson [SS07; Ste10a]. It is inspired by Heisenberg's uncertainty relation and is integrated into a Monte Carlo particle tracing algorithm. Diffraction is applied in the ray casting routine by redirecting particles depending on the proximity to diffracting edges. In contrast to other methods, the algorithm does not directly solve canonical wedge diffraction problems. It is reported sufficient for the employment in room acoustics and noise immission prognosis [Ste96; Ste10b; WSS18], although the stochastic nature of a ray tracing principle contains the risk of missing important paths in a situation where the results are likely to be non-converging.

2.3.1 Maekawa's empirical model

Based on empirical data, a function has been formulated by Maekawa in 1968 for the simplification of diffraction estimation over (and slightly across) barriers inspired by the Kirchhoff approximation [Mae68].

It is elegantly simple and only requires the geometrical detour difference δ over the barrier's edge compared to the virtual line-of-sight distance d to determine an attenuation factor for a selected wavelength λ (i.e., frequency $f = c/\lambda$). The supporting variable $N = 2\delta/\lambda$ (Fresnel zone number) is used and a linear regression curve of the double-logarithmic scale plot resulted in the expression

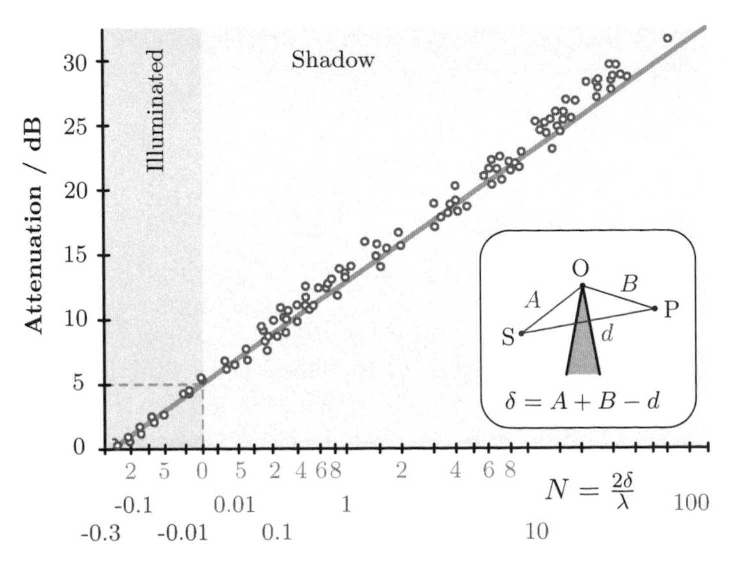

Figure 2.3: Maekawa's empiric data plot with approximate, analytical function extending into the illumination region, taken from Figure 4 in [Mae68].

$$L = 10 \cdot \log_{10}(20N) \tag{2.1}$$

for $N \geq 0$. In the original plot, shown in Figure 2.3, the scaling of the abscissa approaching small values of $N \rightarrow 0.125$ is modified, where the attenuation of 5 dB is theoretically crossed. Maekawa's approximate formula (Eq. 2.1) deviates from the regression curve and fails to settle for the ordinate value of 5 dB at $N = 0$. However, plotting the range of *negative* values for N (which admittedly lacks a physical meaning) in this fashion allows to continue the regression line. This range must be regarded as a description of the detour via the edge in the *illumination area* where source and receiver have a direct line-of-sight and the geometrical connection crosses the edge at grazing incidence with respect to N (which contains the reciprocal wavelength). In other words, Maekawa's plot encourages a case differentiation to account for the predominantly low-frequency diffraction attenuation ($N \rightarrow 0$) at the transition between shadow and illuminated region.

A formula presented by Kurze and Anderson in [KA71] includes this distinctiveness. It is stated that their method offers a good accuracy with a maximum of about 1 dB deviation yet asymptotically approaching the indicated 5 dB constant for the evaluation of $N < 0.125$, yielding

$$L = 5\,\mathrm{dB} + 20 \cdot \log_{10}\left(\frac{\sqrt{2\pi N}}{\tanh\sqrt{2\pi N}}\right)\,\mathrm{dB}\ . \tag{2.2}$$

Because the empirical data was established for a thin screen, the model is generally simplifying barriers to an infinitesimally thin screen, leading to limited accuracy for wide barriers and situations where source or receiver are in close proximity to the barrier surfaces. In addition it must be noted, that the barrier's position and installation angle (which can be defined in a wedge-type model) becomes obsolete as soon as the calculations are reduced to the detour, because this value is independent from the face normals. It is therefore apparent, that urban auralisation applications – apart from thin barrier problems and, to some extent, occlusion-type constellations – will struggle with the integration of Maekawa's model as it is implicitly incapable to model combined diffraction and reflection propagation problems. However, if the geometrical problem is reduced to a detour determination routine (effectively finding a substitution screen of equal detour), approximations like those found in the standards described in Section 2.5 present an engineering method that can handle diffraction for complex geometries.

2.3.2 Uniform Theory of Diffraction (UTD)

The *Uniform Theory of Diffraction* (*UTD*) is mainly characterised by Keller's contribution of light ray diffraction, the *Geometrical Theory of Diffraction* (*GTD*) [Kel62; MPM90], and the work by Kouyoumjian and Pathak regarding electromagnetic waves [KP74]. The formulas to calculate the diffracted sound field are described in Equations 2.3 and 2.4.

Although having its roots in optics, Tsingos et al. integrated the *UTD* into sound modelling of virtual environments using a beam tracing method to identify sequences of diffracting edges [Tsi+01]. The diffraction auralisation is deemed feasible and the importance to combine specular reflections and edge diffractions is emphasised by visualising that energy variations of a dynamic scenario appear generally smooth. The article gives a clear solution how to solve a canonical wedge-type edge diffraction problem mathematically in the appendix, and works out conclusively the importance of separate sound field components for a continuous

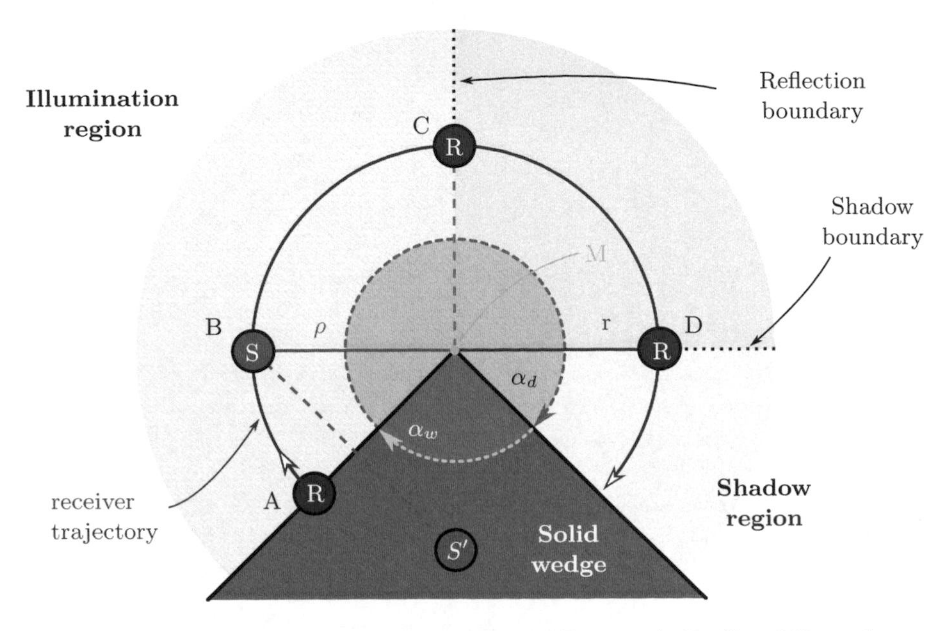

Figure 2.4: Example scene with nomenclature. Letters A, B, C and D are locations on the trajectory of receiver R that correspond to critical transfer functions shown in Fig. 2.5.

sound field modelling. Figure 2.5 partly reproduces the results from [Tsi+01] and demonstrates, how the diffraction component gradually interacts with the direct sound component, until it entirely compensates for the discontinuity occurring at the shadow boundary – where the geometrical line-of-sight abruptly vanishes.

In fact, the Uniform Theory of Diffraction *(UTD)* method is not only capable of handling the illumination-shadow boundary correctly, but blends accurately with the reflection boundary. This is demonstrated by the scene shown in Figure 2.4, where the receiver rotation is extended into the reflection zone. The corresponding acoustic modelling for a selection of frequencies along the extended trajectory is depicted in Figure 2.5. It is chosen in accordance with the scenario of Tsingos et al. in [Tsi+01], Fig. 6a, for comparison reasons.

The scenario shows a right-angled solid wedge as the two-dimensional cross-section. The edge is perpendicular to image plane and the source S and the apex point M are at fixed locations. The receiver R is rotated around the edge from the main wedge face (close to the source, letter A) all the way to the opposite edge

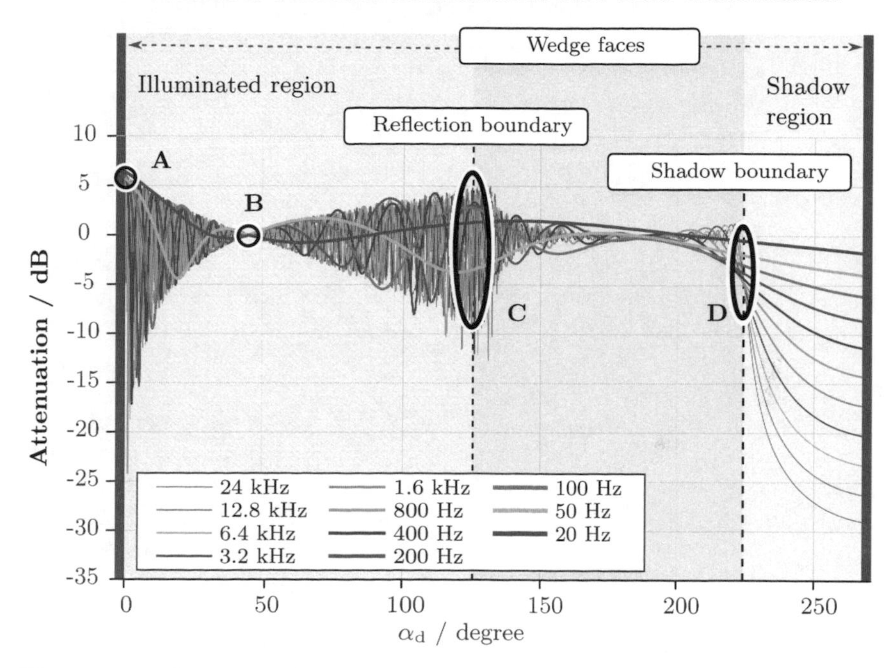

Figure 2.5: Continuous frequency curves of octave band centre frequencies from 20 Hz to 24 kz showing a wedge diffraction scene with a receiver trajectory ranging from one face to the opposite face. Shadow and illumination regions are indicated, the black circles with letter A, B, C and D mark critical boundaries (cf. Fig. 2.4).

face. Areas with different combinations of sound field components are indicated, where the red-inked sections refer to the illuminated region (direct line-of-sight from S to R, i.e., direct sound audible) and the green-inked region shows the shadow zone. Additionally, the deep-red zone to the left of the vertical dashed-line, the reflection boundary, marks the reflection region with respect to the main wedge face.

The corresponding acoustic modelling of the receiver rotation, depicted in Figure 2.5, shows the transfer functions normalized to the respective free-field transmission without the wedge, which is chosen for reasons of comparability with the literature. The curves of octave-band centre frequencies between 20 Hz and 24 kHz start in the illuminated area in front of the main wedge. Although the lines are heavily overlapping due to the superposition of direct and reflected sound fields in the beginning ($\alpha_\mathrm{d} = 0\,..\,180$), the original representative frequencies have been retained, since it is still possible to recognise the continuous development for low frequencies and, at certain locations, the proof of concept remains visible. The black circle at letter A, for example, shows that the circumstances cause an increase in the sound pressure field of approximately +6 dB. This is expected, since the receiver and it's image coincide at $\alpha_\mathrm{d} = 0$ and deliver an in-phase (constructive) superposition with a vanishingly small deviation by the back-scattering contribution of the diffracted sound field component at the edge M. Further, all curves pass 0 dB at the black circle B, where source and receiver are infinitely close making the back-scattering component irrelevant.

At location C, a critical transition is reached where the reflected energy is abruptly removed from the complex-valued equation of geometrically separable sound field components. When paying attention to the low frequencies (20 Hz, 50 Hz, 100 Hz, 200 Hz) it is evident that the diffraction contribution of the UTD model is capable to fully compensate the discontinuity at the critical boundary. This particular feature makes the algorithm highly valuable for the acoustic modelling of dynamic virtual scenarios with reflections. Finally and in consistency with the literature, the compensation of the removal of the direct sound component at location D is shown at the transition into the shadow region.

According to the UTD model, the complex-valued modification of an acoustic incident sound field $H_i(f, M)$ at apex point M can be expressed for a target location R by multiplying a diffraction component $D(f, M)$, a spreading loss factor $A(r, \rho)$ and a phase term e^{-jkr} for the wave number $k = \frac{2\pi f}{c}$ with frequency f and speed of sound c, the distance r from the origin to the apex point and the distance ρ from the apex point to the target yielding

$$H_{d,\mathrm{total}}(f, R) = H_i(f, M) \cdot D(f, M) \cdot A(r, \rho) \cdot e^{-j\frac{2\pi f}{c} \cdot r} \tag{2.3}$$

The complex-valued diffraction coefficient D accounts for amplitude and phase alteration due to diffraction [MPM90; KP74]. It includes all geometrical variables of the edge and relevant relations to the source and target location (cf. Appendix of [Tsi+01] for the definition of $\rho, r, \theta_i, \alpha_i, \alpha_d$ and fraction n):

$$
\begin{aligned}
D(...) = \quad & -\frac{e^{-j\frac{\pi}{4}}}{2n\sqrt{(2k\pi)}\cdot\sin(\theta_i)}\cdot \\
& \Bigg[\cot\left(\frac{\pi+(\alpha_d-\alpha_i)}{2n}\right)\cdot\mathbf{F}\left(kLa^{+}(\alpha_d-\alpha_i)\right)+ \\
& \cot\left(\frac{\pi-(\alpha_d-\alpha_i)}{2n}\right)\cdot\mathbf{F}\left(kLa^{-}(\alpha_d-\alpha_i)\right)+ \\
& \Bigg\{\cot\left(\frac{\pi+(\alpha_d+\alpha_i)}{2n}\right)\cdot\mathbf{F}\left(kLa^{+}(\alpha_d+\alpha_i)\right)\pm \\
& \cot\left(\frac{\pi-(\alpha_d+\alpha_i)}{2n}\right)\cdot\mathbf{F}\left(kLa^{-}(\alpha_d+\alpha_i)\right)\Bigg\}\Bigg]
\end{aligned}
\tag{2.4}
$$

where

$$
L=\frac{\rho r}{\rho+r}\cdot\sin^2\theta_i \tag{2.5}
$$

and

$$
a^{\pm}(\beta)=2\cos^2\frac{2\pi nN^{\pm}-\beta}{2} \tag{2.6}
$$

with $N^{\pm}$ being the integer that satisfies closely the terms

$$
\begin{aligned}
2\pi nN^{+}-\beta &= +\pi \quad \text{and} \\
2\pi nN^{-}-\beta &= -\pi
\end{aligned}
\tag{2.7}
$$

with case differentiations for N^{+},

$$
N^{+}=\begin{cases} 0 & \text{for} \quad \beta\leq\pi\cdot(n-1) \\ 1 & \text{for} \quad \beta>\pi\cdot(n-1) \end{cases} \tag{2.8}
$$

and N^-

$$N^- = \begin{cases} -1 & \text{for} \quad \beta < \pi \cdot (1-n) \\ 0 & \text{for} \quad \pi \cdot (1-n) \leq \beta \leq \pi \cdot (1+n) \\ +1 & \text{for} \quad \beta > \pi \cdot (n-1) \end{cases} \tag{2.9}$$

respectively. The transition function $\mathbf{F}(X)$, initially the integral formula

$$F(X) = 2j\sqrt{X} \cdot e^{iX} \cdot \int_{\sqrt{X}}^{+\infty} e^{-j\tau^2} \, d\tau \tag{2.10}$$

is approximated in the implementation following the suggestion given by Kawai in [Kaw81] for Equation 2.4,

$$F(X) = \begin{cases} \sqrt{\pi X} \cdot \left(1 - \frac{\sqrt{X}}{0.7\sqrt{X}+1.2}\right) \cdot e^{j\frac{\pi}{4}\sqrt{\frac{X}{X+1.4}}} & \text{for} \quad X < 0.8 \\ \left(1 - \frac{0.8}{(X+1.25)^2}\right) \cdot e^{j\frac{\pi}{4}\sqrt{\frac{X}{X+1.4}}} & \text{for} \quad X \geq 0.8\,. \end{cases} \tag{2.11}$$

The smooth transition property for both magnitude *and* phase of the approximate function is depicted in Figure 2.6 for the relevant value range. It is the key to produce smooth transitions at the critical zone boundaries. A complete implementation (source code) of the equations solving a diffraction problem with the UTD method as summarised in [Tsi+01] can be found in the ITA-Toolbox[4].

[4] `http://www.ita-toolbox.org`, see script ita_diffraction_utd.m for MATLAB®

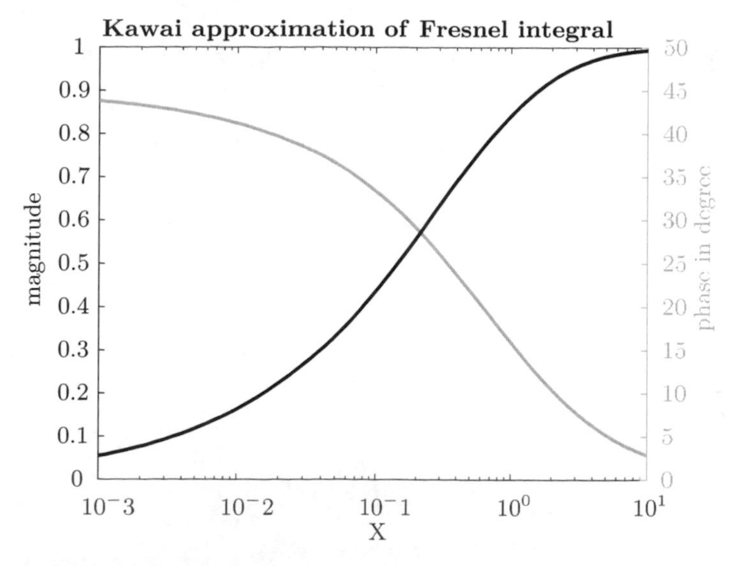

Figure 2.6: Complex-valued Fresnel integral approximation function separated by magnitude and phase for values between $X = 0.001 \dots 10$, after Kawai et al. [Kaw81].

2.4 Environmental noise assessment

The de-facto standard metric for state-of-the-art outdoor noise assessment is the logarithmically scaled mean value of a (frequency-weighted) sound pressure time signal, yielding the scalar sound pressure root-mean-square value

$$p_{\mathrm{rms}} = \sqrt{\frac{1}{T}\int_{0}^{T} p^2(t)\,dt}$$

with T the time period of interest and p the sound pressure as a human perceives it. The decibel representation is determined by

$$L = 20 \cdot \log_{10}\left(\frac{p_{\mathrm{rms}}}{p_0}\right)\,\mathrm{dB}$$

with $p_0 = 20\,\mu\mathrm{Pa}$, the hearing threshold reference value.

Frequency-weighting is carried out by integrating separate frequency bands considered at standardised centre frequencies in the auditory spectrum.[5] A-weighting is the most common approach shifting the band levels with respect to the frequency-dependent human hearing curve. To simplify the quantification of a noise problem to a single value L_A, the formula

$$L_A = 10 \cdot \log_{10} \sum_{i=1}^{N} 10^{\left(\frac{L_i + A_i}{10}\right)}$$

is used, where A_i are the level corrections per frequency band from the literature.[6] Extended perception-oriented methods[7] advance the integration of loudness curves and time-variant masking (see, e.g., [FZ07]). The selection of the integration time T, however, has an essential effect on the final noise level, because outdoor noise emissions are heavily fluctuating. Short-term level values obtained by integrating, for example, one second of a sound pressure time signals are commonly specified as L_S, while averaged values over longer time periods (hours) are referred to as *equivalent* levels[8], L_{eq}. Long-term mean values average out peaks and valleys of short-term samples, however, both representations are relevant in the context of outdoor noise assessment. For this reason, statistical

[5] ANSI S1.11; IEC 225:1966; ISO 266:2003
[6] IEC 61672-1
[7] ISO 12913-1:2014
[8] DIN 45641

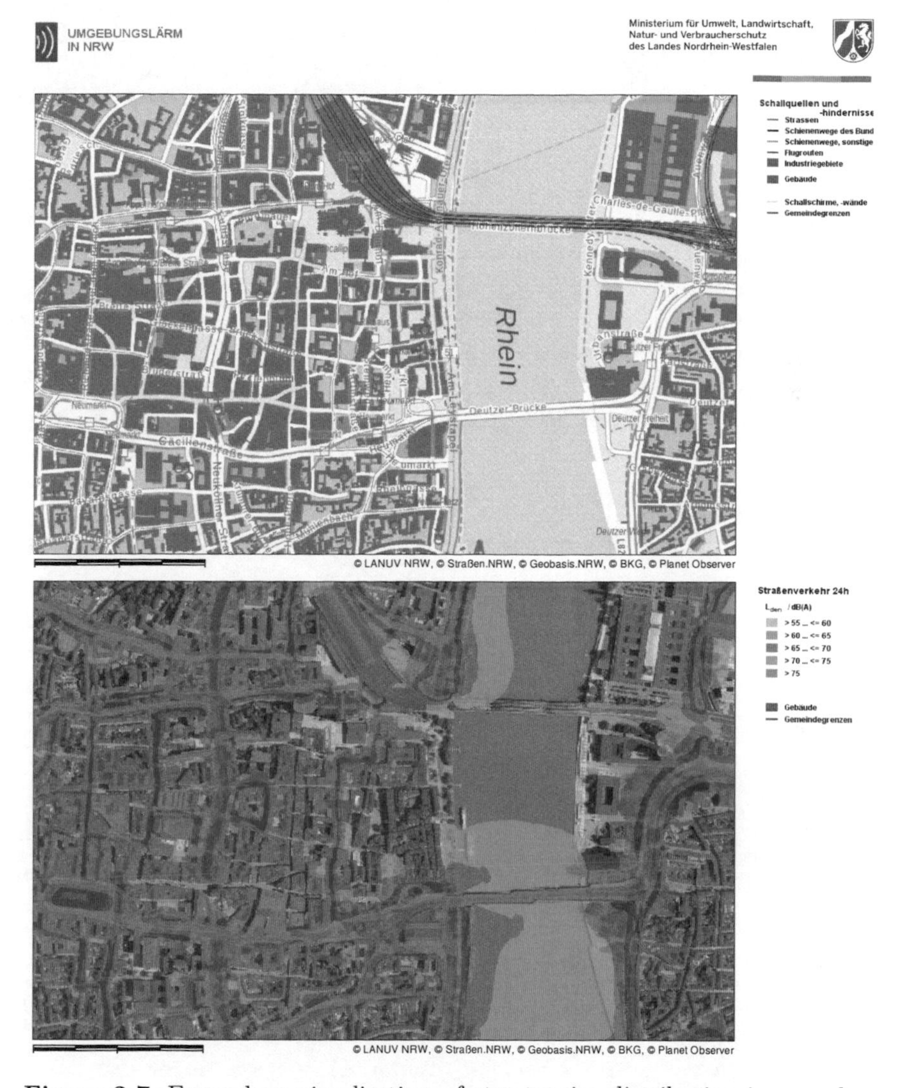

Figure 2.7: Exemplary visualisation of street noise distribution in an urban environment, delivered by the open data service of North Rhine-Westphalia's environment department.
Umgebungslärm in NRW, Ministerium für Umweltschutz, Landwirtschaft, Natur- und Verbraucherschutz des Landes Nordrhein-Westfalen, exported March 2021 from `https://www.umgebungslaerm-kartierung.nrw.de`

evaluations of short-term measurement data series are employed to maintain information on the fluctuation. Especially the percentages of the high-level noise events are of interest as they reflect potentially disturbing exposures that can lead to an increased health risk, for example, during night-time [Wor11].

The presentation of noise assessment reports include either tabulated noise levels or visualisations in form of noise contour maps. State-of-the-art online platforms simplify the access and improve the comprehension of urban noise distribution significantly. For example, the free noise map system[9], an open government initiative provided by the Ministry for Environment, Agriculture, Conservation and Consumer Protection of the State of North Rhine-Westphalia, Germany, specifically targets citizens and uses open data based on the geospatial standard Geographic Information System *(GIS)*[10]. It has an interactive component making it possible to switch between views, layers of day and night-time levels, and distinguished contributions of noise source categories. An export of an urban environment is depicted in Figure 2.7 showing a) the geometrical map with the lines of noise sources differentiating between streets, tram routes and railways, and b) the corresponding street noise contour map with absolute decibel values averaged over 24 hours, L_{DEN} (day, evening, night). Noise level contour maps are simulated in accordance with noise prediction models (in this case the German RLS-90 norm[11]). They will be reviewed briefly in Section 2.5. Traffic is composed of countless vehicles moving at various speeds that represents a short-term event from a noise immission perspective. However, noise simulation methods are founded on statistical values of traffic volume and speed limits, which itself are based on prediction models that integrate, for example, empiric sampling at representative locations. In addition, weather conditions impact the sound generation and propagation attenuation in urban environments, which is a factor that once again must be averaged out in order to maintain a robust way to describe general noise levels.

This kind of noise assessment helps to identify potentially loud areas and can be consulted to estimate the expected noise immission at a certain location in the city. However, due to the averaging in various dimensions, the delivery of statistical noise levels lacks a deeper comprehension of the soundscape on a micro-time scale. Although the information can be partly extracted from the statistical analysis (e.g., by considering the accumulated peak values in a measurement),

---

[9] MULNV NRW, Germany, `http://www.umgebungslaerm.nrw.de`

[10] `https://www.ogc.org`

[11] Richtlinien für den Lärmschutz an Straßen, Verkehrslärmschutzverordnung, 16. BImSchV (German legislation)

interpreting and quantifying these values requires a higher level of understanding and largely excludes non-specialists.

Real-time auralisation can deliver both an intuitive interface to grasp the expected soundscape and allows an interactive personal evaluation of the noise composition – an appealing consultation method and convincing way to present noise aspects with regards to a decision process that involves stakeholders of different disciplines. However, such an attempt must overcome many limitations and, as of today, must be seen at most as a desirable feature extending state-of-the-art outdoor noise assessment methods. Put simple, the auralisation of urban environments with the currently available data is able to integrate realistic weather scenarios, can make use of random processes and traffic flow simulations to emulate moving vehicles on streets, on railways and in the air. As will be described, the built environment can be integrated to a certain extent into noise propagation methods and, considering that there is an authentic way to synthesise a characteristic sound of a noise source, it is possible to create a computer-generated virtual urban environment that can be perceived naturally by the human (binaural) auditory system.

2.5 Standards

To effectively realise a legally binding environmental noise assessment and management directives, standards must be defined that clearly and unambiguously define the principles, models and approximations to be met in a judicial area.

Several national and international methods and standards have been formulated by committees and institutions like the International Organization for Standardization *(ISO)* and the European Union *(EU)*. The content of these documents describe in detail the compliance requirement of measurement equipment, the list of accredited procedures to conduct measurements, to determine metrics and to predict values by simulation, and finally stipulate valid ranges and restrictions in form of directives and regulations.

To narrow down this complex apparatus and bring the essential parts into focus, the technical standard *ISO 9613* from 1986 and the *CNOSSOS-EU* method from 2015 are briefly reviewed in Section 2.5.1 and Section 2.5.2, respectively. Relevant aspects deemed applicable to the subject of outdoor sound auralisation are highlighted while less important sections are omitted.

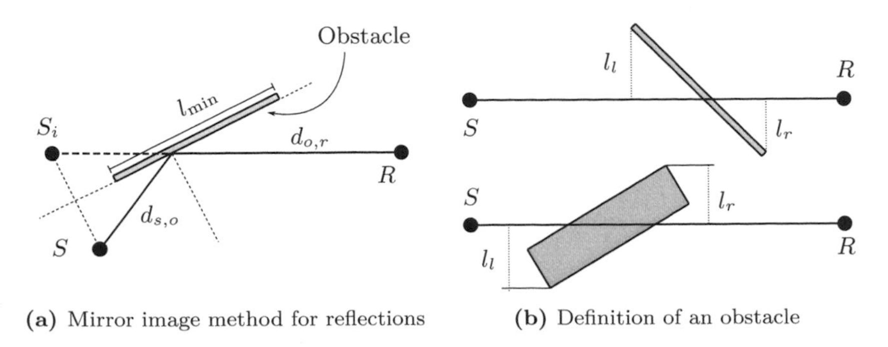

(a) Mirror image method for reflections

(b) Definition of an obstacle

Figure 2.8: Propagation handling for geometries described in ISO 9613-2 (cf. [ISO96] Fig 4 and Fig 8).

Both approaches essentially describe the sound propagation of individual sound sources in a residential area comprising stationary noise locations, traffic lines and buildings. Compared with current technological possibilities, ISO 9613 describes a rudimentary engineering method that employs many simplified models, which do not pose a problem for today's computers. However, the initial idea to consider sound sources separately and determine attenuation values as a consequence of sound propagation forms the basis of noise assessment prediction as applied today. The *Harmonoise* project [Def+07; MD07; Sal+11] formulated methods for strategic noise maps that finally led to to the Directive 2015/996 establishing common noise assessment methods in the EU, referred to as *CNOSSOS-EU*. It is particularly interesting as the methodology harmonised many other methods by integrating, for example, parts of ISO 9613, the Scandinavian environmental noise prediction method (Nord2000) [PK06] and the French road noise prediction (NMPB2008) [Kep+14; BD09].

2.5.1 ISO 9613

The ISO 9613 standard from 1986 is the most established method to predict noise outdoors. In two parts, it describes general outdoor sound propagation calculations to determine equivalent A-weighted sound pressure levels (cf. ISO 1996) at a receiver location for a selection of source types [ISO93; ISO96].

The first part specifies analytical formulae to calculate absorption by the atmosphere as a function of frequency. Atmospheric conditions can be parametrised by temperature, relative humidity and static pressure. Annex C informs about

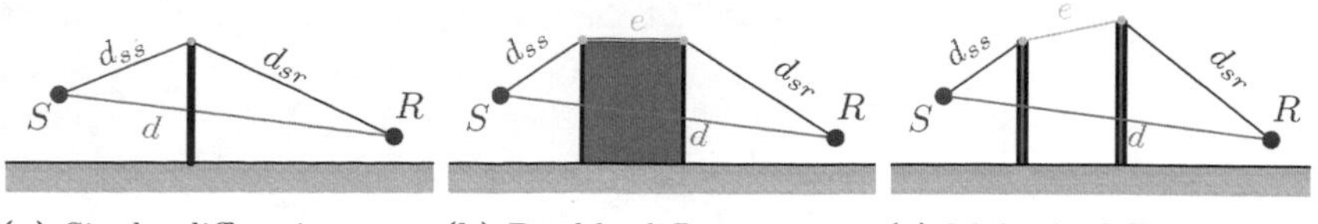

(a) Single diffraction at a screen
(b) Double diffraction at a wide obstacle
(c) Multiple diffractions at cascaded obstacles

Figure 2.9: Determination of detour for obstacles in the cross-section of a vertical plane as described in ISO 9613-2 (cf. [ISO96] Fig 6 and Fig 7).

approaches for an altitude-dependent procedure that is applicable for vertically inhomogeneous meteorological environments, like a stratified atmosphere. Furthermore, Annexes D and E provide instructions and examples for handling spectrum integration and A-weighting compliant attenuation determination after IEC 651.

The second part, the *general method of calculation*, presents an engineering approach to predict sound levels at a distance. It includes aspects of atmospheric absorption, wind conditions, geometrical diversion, ground effect, reflections from surfaces and screening by obstacles. Additionally, information is provided how to apply correction factors concerning propagation through a rudimentary built environment, foliage (e.g., trees) and industrial sites.

Figure 2.8 visualises the integration of geometry into the propagation algorithm. An obstacle is subject to reflection or diffraction, if the spatial dimension is larger than the respective wave lengths. While reflections are handled according to the classical mirror image model [AB79], diffracted paths are approximated by projecting the propagation problem into two-dimensional space on horizontal and vertical planes, and evaluating the convex hull of the intersection, as depicted in Figure 2.9. This way, the detour value can be determined for single and multiple diffractions as well as cascaded screens.

Regarding auralisation, the calculation of sound propagation over distance including atmospheric attenuation is applicable. In particular, the differentiation between sound sources that generate noise and environmental propagation effects that accumulate or reduce the transmitted acoustic energy at the receiver location is equivalent (cf. Section 2). However, auralisation requires operations at higher resolution. Firstly, the energy-based source sound power band spectrum must be substituted with a time signal. Secondly, propagation simulation must explicitly solve reflections with a coherent model and requires comprehensive mechanisms to integrate diffraction that exceeds the engineering methods described in ISO 9613-2. Lastly, A-weighting can be omitted because the auralisation result is perceived by the human's auditory system. The reduction of the receiver to a virtual probe that does not maintain directional cues does not suffice a comprehensive auralisation output and must be expanded by a perception-equivalent model (e.g., a binaural processor, cf. Section 4.7.2) or an acoustic reproduction system that maintains perceptual cues (such as Vector-Base Amplitude Panning *(VBAP)* [Pul97], Higher-Order Ambisonics *(HOA)* [Ger83] or Wave Field Synthesis *(WFS)* [BVV93], cf. also Section 4.8).

2.5.2 CNOSSOS-EU

The *CNOSSOS-EU* commission directive from 2015 describes *Common Noise Assessment Methods in the EU.* It refines earlier directives that are intended to harmonise approaches for all Member States (of the EU). Apart from a regulatory body regarding the establishment and verification of strategic noise maps using indicators such as L_{den} and L_{night}, a methodological framework has been developed that provides guidelines for the calculation of noise levels at receiver locations from various sources including specifics of sound propagation [Sal+11]. *CNOSSOS-EU* has been published as *Reference Report* by a Joint Research Centre.[12] The report includes an extensive database that delivers tabulated values describing noise sources and correction factors required by the calculation methods in order to account for different source-related conditions (e.g., vehicle speed and road surfaces). Similarly to ISO 9613-2, a separation of sound sources and propagation effects is proposed. Emitted sound power is defined in frequency bands including a directivity property, the working hours (day, evening, night) and an operating condition, if applicable. A source object is attributed a class that discriminates point, line and area type, and the properties of location, orientation and dimension are assigned. Calculation methods of noise propagation differentiate between ground sources (road, railway and industrial sources) and elevated sources (aircraft sources). Ground sources can be based on two particular types of atmospheric conditions, in a homogeneous medium (including the mean *favourable condition*) and under downward-refracting propagation conditions from the source to the receiver. Specular reflections are considered by a mirror image model, screening is approximated as a single diffraction calculation, and a method for two or more screens is described. Relative to the receiver lo-

[12] S. Kephalopoulos, M. Paviotti, and F. Anfosso-Lédée, Common Noise Assessment Methods in Europe (CNOSSOS-EU), JRC Reference Report, EUR 25379 EN. Luxembourg: Publications Office of the European Union, 2012

cation, a segmentation of sound sources is proposed depending on distance, and superposition is performed incoherently. Generally, the CNOSSOS propagation method operates on a geometrical model that consists of a connected mesh representing ground surface and *obstacle surfaces* (i.e., buildings, topology, barriers that present a certain slope with respect to the vertical axis). The employment of a pathfinder engine is encouraged. However, propagation paths are not determined in the original space, but are projected using vertical and horizontal planes containing the corresponding source and receiver, and are interpreted in the sense of a detour approximating Fermat's principle. Significantly large topological heights, like hills or road embankments, are considered separately and are effectively used to replace the otherwise horizontal ground (mean plane and equivalent heights). The contribution of each individual propagation path to the noise level at the receiver location is taken into account by applying a sound propagation model with averaging of different conditions over time to obtain long-term sound levels. The total levels are determined by energetic summation of all paths between source and receiver. The consideration of geometrical divergence, atmospheric attenuation and boundary interactions like ground effect and screening is applied by subtracting the spectra from the initial sound power spectrum in the same fashion as described in ISO 9613. However, a more elaborate approximation of the general ground effect is given and a detailed attenuation calculation is described in case of diffraction along the direct path that also affects the attenuation of the reflected path via ground. In principle, an engineering approach is provided that determines an attenuation spectrum based on wavelength plus the Euclidean detour value via the top of the obstacle (leading back to Maekawa's model [Mae68], cf. also Section 2.3.1). The method takes into consideration the simplified geometrical propagation paths from the source via the equivalent diffraction point with the ground reflection on both source and receiver side regarding the equivalent heights, as depicted in Figure 2.10 for a single diffraction problem.

Multiple diffractions are integrated by finding the shortest geometrical path via all obstacles imagining a convex hull covering the geometry. However, the method is considering the intersection by a vertical plane through source and receiver, effectively projecting the problem from a three-dimensional to a two-dimensional space. The report notes that reflections *and* diffractions bear a relationship that influences the (combined) attenuation using the terminology *retrodiffraction*. Solutions are proposed to take measures in cases like trenches, where acoustic energy only reaches the receiver after multiple reflections via the same faces. It is argued that the method aims to ensure field continuity in the transition area towards the border where the direct sound path gains relevance

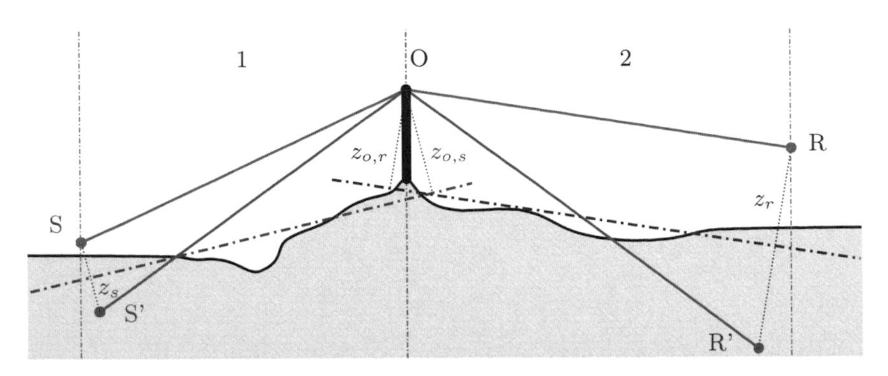

Figure 2.10: Determination of geometrical diffraction and reflection relations over non-planar ground after CNOSSOS-EU Reference Report, Fig. 2.5.c.

(using the terminology clear area and shadow area). Further, it is mentioned that combinations of reflections and diffractions can occur and a substitution operation is suggested to resolve image sources and image receivers effectively simplifying the problem for paths that start or end with a specular reflection and potentially include one or many diffractions in-between. Again, the method suggests to apply the retrodiffraction correction for these substitute, equivalent paths.

An optional statistical method in an urban area is envisaged that allows to account for sound propagation corrections after the first line of buildings. However, no specific instructions are given, except that the method must be duly documented, if applied, and must be based on average density and height of the built environment.

The second part of the framework regarding elevated sources is mostly concerned with the definition of sound sources under different conditions, meteorological influences and the specification of time-averaged single-value metrics (e.g., exposure levels). For the most part, the report covers aircraft noise, which is especially complicated, since the noise pollution is mainly characterised by distinct yet highly fluctuating fly-over noise events. Therefore, meticulous efforts have been put into the handling of flight plans, approach and departure of aircrafts, radar data evaluation and flight track / flight profile interpolation, and many other aspects. In the *context of auralisation*, which by nature considers acoustic events like a fly-over individually, the provided information is relevant mostly for the source description. The propagation component through the atmosphere takes a secondary role in the vicinity of buildings in an urban setting. However, noise from elevated sources can be highly relevant in certain scenarios and it appears relatively straight-forward to combine the otherwise clearly separable components of propagation through the atmosphere and the transmission via an urban scenario. For this reason, particularities of aircraft noise and the handling of sound propagation through the (inhomogeneous) atmosphere are not discussed.

2.6 Simulation and auralisation

To implement auralisation for virtual outdoor environments, the scene must be separated into sources, receivers and environmental data, like a geometry mesh and meteorological conditions, as will be described in detail in the following (cf. Chapter 4). A rendering engine is required that processes each emitted source signal and applies the individually simulated propagation footprint by means of DSP [Vor11]. Additionally, a method to physically reproduce the rendering output by headphones or loudspeakers is obligatory. Depending on the application, the focus can be placed on the source characteristics or on the sound propagation effects. Also, if an interactive feature is desired, real-time processing must be integrated. This is particularly important in the context of VR applications and the entertainment sector that employs audio engines in games.

In the past, scientific projects have investigated methods to realise outdoor sound auralisations that were mostly motivated to reflect physical-based behaviour as accurately as possible for a given scenario. The approaches were either based on prototypes, used more versatile auralisation frameworks or employed audio plugins provided by commercial VR environments. A selection of relevant work from the literature is reviewed and discussed in the context of outdoor sound auralisation.

As early as 1983, Moore describes a general model to produce spatial sound for a virtual space by considering the human auditory system [Moo83]. He points out the need for physics-based computer simulation and perception-related spatialisation, and anticipates a compromise between psychophysical and performance matters. In other words, implementing a virtual acoustic environment must aim to present a *reasonable* illusion of the space with the resources at hand. A network of DSP units is proposed to cover distance cues and Inter-aural Time Difference *(ITD)* by several time-varying tapped delay lines (i.e., Finite Impulse Response *(FIR)* filters with dynamic coefficient exchange) that apply a frequency-dependent gain, and several recursive filters to simulate room reverberation (Infinite Impulse Response *(IIR)* filters). The reproduction system was inseparably integrated into the approach and follows a panning principle via multiple loudspeakers rather than a binaural approach for headphones (see, e.g., [Bla97; Bla+00]).

In 1993, Kleiner, Dalenbäck and Svensson published an auralisation overview paper that nicely categorises and summarises the history of auralisation, which was predominantly driven by the room acoustics community [KDS93]. A more general auralisation approach is described by Lokki in [Lok+02]. It presents the latest progress of the *DIVA* system that realises a visionary sound rendering pipeline by Takala et al. [TH92; Sav+99; LHS01]. It is mentioned, that the dynamic simulation of propagation including reflections and diffractions, which is a feature that has the potential to also be transferred to outdoor scenarios, has been based on methods published by Pulkki et al. in [PLS02] (cf. Chapter 3). In particular, the separate handling of *individual* early reflections with a Fractional Delay *(FD)* filtering unit and a subsequent directional filtering based on Head-Related Transfer Functions *(HRTFs)* provides substantial advantages for outdoor sound auralisation applications, as will be addressed in Chapter 4. However, most applications seemed to either consider static sources, use pre-calculated simulations or provide a dynamic feature limited to head movements of the user, which only affects the spatialisation processing.

Another approach that was based on convolution and filter exchange combined with a cross-fading routine to account for a dynamic environment – including source movement – was published by Tsingos and Gascuel [TG97]. The method mentions the applicability of a variable delay processor using resampling to account for time-variant Doppler shifts based on a time model for retarded movement.

In 2000, Wenzel et al. published the Sound Lab real-time auralisation framework for the study of spatial hearing [WMA00], which is noteworthy, because it does not initially target room acoustics and instead focuses on a solid realisation for general real-time operability. The design of the system includes many modern aspects, like an application programming interface and a clear distinction between input handling and update routines. Signal processing parameters are made available by a physical simulation layer and signal processing units are rendering source-receiver pairs separately, referred to as *auralisation units*. In particular, the concept includes a functional layer that improves the input values in order to provide an adequate parameter stream for the auralisation units, for example, by smoothing motion trajectories.

An implementation of a scalable generator providing the auditory rendering for the use in a multi-modal virtual environment is presented by Blauert et al. in 2000 [Bla+00]. The method is generally applicable for outdoor scenarios but was intended for real-time room acoustic auralisations. A purely binaural approach with dedicated hardware components for high interactivity is described that processes individual geometrical propagation paths with a *(binaural) auralisation unit*. Each unit uses one delay module, several filtering stages and a two-channel binaural FIR processing component for the HRTF. In particular, the necessity for implicit time-variant processing of fast moving sound sources to incorporate Doppler shifts is mentioned, and a solution is proposed by integrating methods of Strauss [Str98].

In contrast, a similar architecture realised in software is represented by Silzle et al. with the IKA-SIM system [SSN04]. It explicitly manages dynamic sound field components like the Doppler shift by time-variant processors using length-modulated digital delay lines (i.e., interpolating Variable Delay Lines *(VDLs)*, cf. Section 4.6). The concept is built on a per-path rendering pipeline with a clear separation of delay routine, spectral filtering and directional processing that implements spatialisation either for headphones using binaural technology, or a panning method for loudspeakers. Interactive rates are reported to be achieved for moderately complex geometries with an accelerated low-order mirror image method and the respective DSP layout [Leh93], which is remarkable considering the available computer architectures at that time.

In literature, contributions on sound rendering also emerged from groups that are predominantly concerned with computer graphics (e.g., visualisation, visual rendering and Augmented/Virtual Reality) and transferred established visualisation methods to the acoustics domain in order to provide convincing sound for games. For example, the *RESound* engine presented by Taylor et al. aims at providing an audio feedback for interactive game-like environments by utilising volumetric ray-tracing to determine propagation paths and derive the composition of early and late reverberation [Tay+09b; LCM07; Tay+09a]. A clear focus on the algorithmic implementation and acceleration methods using spatial data structures to determine geometrical propagation paths is noticeable. The actual rendering is divided into several filter units that include directional processing for first-order paths and reduce reverberation to a non-directional band-selective filtering procedure that uses cross-fading and interpolation components to smooth the output on parameter change. However, comprehensive time-variant digital signal processing methods with parameter interpolation for individual paths regarding dynamic scenes have been introduced later [Tay+12]. Further work by

Schissler et al. reports on the integration of high-order diffractions and diffuse reflections, and propose convolution methods for handling complex multi-source environments, that are feasible for outdoor scenarios [SMM14; SM16b; SM16a].

Another path is followed by Rungta et al. in [Run+16], where the classical separation of sound source and medium propagation is revised in order to render a coherent radiation pattern that is not assuming point sources.

Mehra et al. present a wave-based auralisation approach for large outdoor scenarios that makes use of pre-calculations to represent sound sources as equivalent objects that interact with geometrical objects of the environment [Meh+12; Yeh+13]. A summation for a listener position takes source emission and sound propagation into consideration at interactive rates and is proposed for noise prediction applications, although dynamic sources are not covered.

Building on the same idea, Raghuvanshi et al. propose similar methods known as light-baking from visualisation for acoustic parameters throughout a virtual environment [RS14; RS18]. Pre-calculated compressed data samples at locations where sources and receivers are likely to reside are used to interpolate and mix a convincing dynamic audio output that approximates various relevant propagation effects, for example, a reverberation decay and occlusion depending on the scene geometry. Particular efforts are made to preserve a continuous loudness development that avoids glitches when visibilities of important individual propagation paths are toggling.

Perception-based acceleration methods have been proposed that organise and simplify highly complex acoustic scenarios, for example, by clustering sound sources in order to facilitate the simulation effort and the computational load on the signal processing chain [TGD04; Tsi07; Moe+07; FBH97]. The exploitation of findings from psycho-acoustics, auditory scene analysis and perceptual audio-coding is promising, however, can be costly as the approaches require perceptual models that evaluate the scenario in real-time in order to apply simplification measures. Additionally, progressive methods are susceptible to cause perceivable effects for unusual scenarios and require careful validation. Nonetheless, with currently available resources, the integration of auditory culling and spatial level-of-detail methods are unavoidable for a comprehensive real-time auralisation when applied to urban scenarios. In particular, the work by Tsingos et al. explicitly tackles challenges of outdoor environments [TGD04] containing many sound sources. For example, a dynamic grouping routine based on a distance metric is proposed to generate *pre-mixed* signals for *impostor* sound sources, effectively decreasing the load on propagation simulation and rendering (see also [Her99] for moving sources). Also, it is suggested to interactively cull sources that are subject to perceptual masking considering an estimation of loudness contribution at the receiver after applying fundamental propagation attenuation. Because the auralisation approach uses an image source model, reflections can be regarded as secondary sources and are handled equivalently, for instance, the influence of individual geometrical paths is evaluated for the culling procedure. A compromise between acoustic quality and computational effort must be met, which is founded on psycho-acoustic evidence that auditory perception is limited regarding the localisation capability [Moo83; Bla97].

Scheduling methods, as described by Fouad et al., support the real-time capacity of sound engines by making use of importance sampling and interpolation where update rates become insufficient [FBH97; FNH99; FBB00]. In other words, such resolution-scaling algorithms can be adapted to distribute the limited resources and focus on relevant parts of various functional layers, such as simulation, modelling and signal processing. However, the estimation of an *importance metric* (e.g., attention) is subjective for a listener and prediction models are highly context-dependent making it impossible to suggest a generally applicable approach.

In 2012, Manyoky at al. presented a participatory audio-visual instrument called *VisASim* to improve social acceptance of wind power planning [Man+12]. Validation of the prototype based on a game engine was conducted by comparison with recordings. Follow up publications describe in more detail the auralisation approach, which separates between emission model and propagation model, as well as the reproduction system. The synthesis of source signals is basically founded on a sophisticated analysis-and-synthesis approach with an option to adjust property and state of the wind turbine, as described by Pieren et al. in [Pie+14]. After Heutschi et al., the subsequent propagation simulation establishes a short-length convolution parametrised in the frequency domain with complex models accounting for atmospheric fluctuations over time, since moderate to strong wind conditions are expected during operation of wind turbines [Heu+14]. In the same paper, a vegetation source signal generator using spectral noise shaping is presented that models, among other aspects, rustling leaves of given foliage properties modulated by a time-variant wind speed parameter. This type of environmental noise can be a complementary feature for urban scenarios in order to increase realism or investigate masking effects relevant, for example, in a park situation. The described approach has been transferred to other types of outdoor noise events, namely accelerating car pass-bys [PBH15; PBH16], aircraft fly-overs [Pie+19] and railway noise auralisation [Pie+16] by substituting the dynamic signal synthesis component and adapting the propagation model, for example, by integrating Doppler shifts using resampling for fast-moving sources. The available publications and multimedia contents suggest that the prototype auralisation application has not been employed in interactive VR systems with the option to move freely, yet.

Many more examples exist that implement engineering methods to provide auralisations for specific purposes using different sound propagation models, for example, in the context of car pass-by sound event rendering under various conditions [GHK19; Pep+11; Tho+12; MJ13; Jol+15; GHK19].

A conclusive auralisation prototype for outdoor sound is proposed by Viggen et al. in [VVO+15] that consequently builds on the propagation methods described in the Scandinavian Nord2000 noise prediction standard (cf. Section 2.5) and aims to produce a HOA audio stream that preserves the freedom to rotate an interactive listener (cf., e.g., [Raf04]).

A different approach to the auralisation of a pass-by vehicle event is described by Georgiou et al. [GHK19; Men+18]. In contrast to most realisations that employ GA models, binaural propagation Impulse Responses *(IRs)* at distinct locations sampled along a pre-determined trajectory were first measured and then simulated by a two-dimensional, pseudo-spectral, time-domain method for low frequencies. Synthesised car signals were convolved and mixed with a sufficiently long cross-fading window, which was validated by a listening experiment. However, tonal components in the source signal are reported to result in unnatural auralisations (cf. Section 4.6).

Another comprehensive auralisation application developed in the context of hearing aid and audiology research is presented as an open-source toolbox by Grimm et al. in [GLH19]. The TASCAR system can be potentially used for outdoor sound auralisation as it covers specular reflections and features dynamic source and receiver movements claiming to process several hundred entities on commodity hardware. This is achieved by extremely short, per-path filtering in the time-domain (up to second-order low-pass FIR filters) to account for frequency-dependent attenuation and a time-variant delay line with nearest-neighbour read-out to model the propagation delay. In particular, time-variances are incorporated and modifications of the geometry are possible, which stands out in comparison to other approaches. However, the missing integration of diffracted sound components limits the feasibility for urban noise assessment.

The approach described in this work is based on Virtual Acoustics *(VA)*[13], an open-source real-time auralisation framework initiated, continuously improved and extended by the *Institute of Technical Acoustics (ITA)* at *RWTH Aachen University*, Germany [SV19b]. Initially, the project aimed at generating room acoustics for virtual environments with dynamic elements. An early version simulated direct sound and early reflections in the binaural rendering processor with real-time update rates and employed a look-up procedure for pre-calculated (and later also real-time simulated) reverberation filters using the software *RAVEN* [Sch+10; Pel+14; APV14]. Also, a dynamic, non-intrusive, transaural reproduction method was described by Lentz et al. based on a cross-talk cancellation module making the auralisation system particularly feasible for *CAVE*-like environments with many screens and inconvenient options to place loudspeakers [Cru+92; Len06; Len+07]. Later, Sahai and Wefers reconditioned the framework and extended the functionality for aircraft noise evaluation in a real-time VR application [Sah+12]. In particular, a new time-variant handling of Doppler shifts was required and efforts were made to improve the refinement of asynchronous motion data (i.e., under-sampled and time-distorted trajectories) [Sah+16; WV14; Wef17; WV18]. A scheduling functionality was introduced to enable dynamic room acoustic simulation for direct sound, early reflections *and* late reverberation at different refresh rates [Wef+14]. The approach implements a scalable back-end architecture for distributed computing and hybrid parallelisation. In the process of this work, concepts, modules and features for outdoor sound auralisation were added, both for offline rendering and real-time applications [VS15; SV15; SAV19]. In particular, progress on the integration of diffracted propagation paths into the simulation stage with an explicit suitability for real-time auralisation is described by Stienen et al. [SV17] and Erraji et al. [ESV21]. At the same time, modern layouts to render individual propagation paths are described, that take into consideration distinctive outdoor acoustic phenomena and produce a perception-oriented output using binaural technology [SV18; VAS19; Asp+19].

[13] `http://www.virtualacoustics.org`

3 Real-time auralisation of outdoor sound

To make a dynamic outdoor environment audible by means of auralisation, a concept meeting real-time prerequisites is required. In particular, the fast determination of acoustic propagation based on a computationally demanding physics-based simulation represents a major challenge.

The associated simulation techniques can be generally classified into wave-based methods and geometrical methods (or ray-based methods) [KDS93]. In the context of real-time auralisation, Geometrical Acoustics (GA) methods have been given priority for most systems and frameworks described in the literature due to their relative simplicity and suitability for a point-to-point sound propagation problem (cf. Section 2.6). Consequently, special attention is paid to *fast* methods using GA principles. Two own implementations dedicated to urban environments with a comprehensive feature to address diffracted paths are described in Section 3.6 and Section 3.7. Since the integration of *diffraction models* into geometrical approaches is of particular importance for urban environments, the feasibility of available approaches is discussed in Section 3.5.

3.1 Auralisation concept for dynamic environments

The implementation of (real-time) auralisation engines puts high demands on the sound propagation simulation procedure, necessitating highly efficient algorithms in terms of computational runtime and resource consumption. Of equal importance is the quality of the result, which must suffice the desired physical and perceptual aspects. Algorithms based on the GA principle stand out in these categories as they are generally lightweight and have the capability of modelling the relevant acoustic phenomena delivering sufficiently precise results in the audible frequency range, which represents a compromise between plausibility and accuracy [KDS93; Beg00; Vor11].

In the literature, various auralisation approaches have been presented that were developed for specific purposes, see Section 2.6. Different frameworks with the potential to be feasible for *outdoor sound auralisation* regarding noise assessment at interactive rates apply GA propagation models in the simulation stage. Additionally, dynamic characteristics of a scenario are reflected by time-variant signal processing routines in the rendering stage, which commonly includes a per-path routine and requires a separation of simulation output parameters into a delay, an attenuation spectrum and directional information at the receiver location. Regarding real-time capability, most described systems are reported to be generally fit to produce updates at interactive rates. However, it is often unclear under which limitations real-time auralisation is accomplished. Although it is difficult to generalise the constraints, real-time applicability is commonly achieved assuming a single receiver and a limited number of sound sources which can be processed at low-latency audio frame rates. Moreover, the simulation result rates are effectively increased by reducing the resolution of the propagation models, most prominently by limiting the order of image sources to a low number with underlying reasoning that human perception is likely to be less sensitive in interactive auralisation, even more so in audio-video applications. Last but not least, the incoming stream of simulation results from a comprehensive geometry-aware algorithm under-samples the required rate for a seamless time-variant audio rendering, which is necessary for the implementation of the Doppler shift. This mismatch is resolved by interpolation routines that produce intermediate parameters and effectively allow for a smooth fading between simulation frames. Every aspect of the enumerated limitations requires a perceptual validation tailored to the given context. Particularly for outdoor scenarios, a compromise between accuracy and efficiency becomes apparent, as a decision must be taken whether many sources with low-resolution simulation feeds are preferred over a limited source count with a higher simulation resolution. However, the final capacity of propagation paths that can be handled is independent from the simulation quality and shifts the investigation towards the question, which paths are important and need to be prioritised, a problem that is addressed in Section 4.5 and discussed in Chapter 6.

Considering these aspects, an outdoor sound auralisation concept is derived and illustrated in Figure 3.1. The concept differentiates between three contexts: The *user context* describes the application layer that controls the outdoor environment that, for example, transmits motion data from a MATLAB® or Python script, or a VR designer like Unity®[14] and Unreal®[15]. In the *auralisation context*, an auralisation unit manages the virtual scene with the dynamic objects, provides a

[14] Unity Technologies, `http://www.unity3d.com`
[15] Epic Games, Inc., `http://www.unrealengine.com`

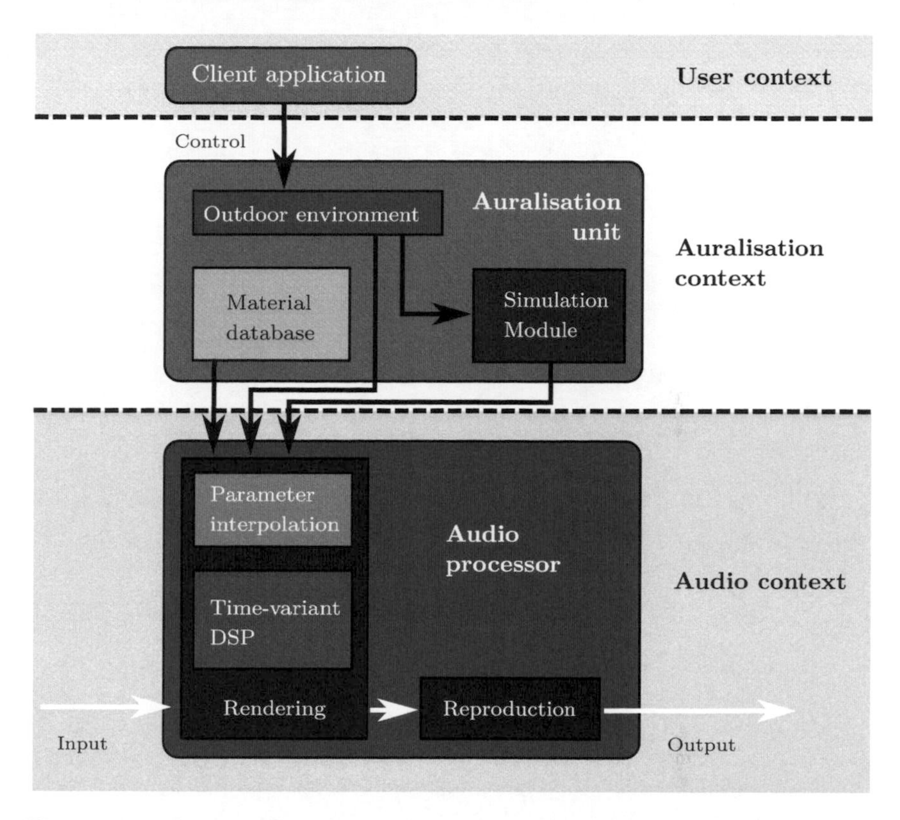

Figure 3.1: Concept for a highly-dynamic outdoor sound auralisation system.

material database and controls a simulation module that produces propagation results. The last layer, the *audio context*, processes digital audio streams by means of DSP and modifies source signals (input streams) in a way that the virtual environment is acoustically rendered. Since the audio processing is performed at significantly higher update rates than propagation simulation data can be provided, an interpolation module is needed to estimate intermediate values, such as motion trajectories, in order to obtain a plausible auralisation result.

This concept is appropriate for both offline auralisation and real-time processing. The fundamental difference is that the audio context can be artificially suspended by an arbitrary amount of time, if no real-time constraints apply. This is helpful to improve the overall quality of an auralisation at the cost of a longer processing time. For example, a procedure could wait for simulation results before continuing with the audio rendering. Hereby, the propagation simulation update rate can be synchronised with the audio processing, at the cost of an increased simulation effort, making the interpolation of results obsolete. Also, an arbitrarily high number of sound paths can be rendered, as the time budget for the DSP network becomes irrelevant. In contrast, a real-time audio callback process from a real audio device expects that the output stream is provided within a given time frame. If not delivered in time, dropouts occur and the system is not operating as expected. As a consequence, a challenging restriction is imposed on the DSP network, making it necessary to introduce a maximum capacity of paths that are processable in real-time. Since it is impossible for a propagation simulation engine to keep up with the audio frame rate, simulation results are produced at much lower rates and require parameter interpolation, if perceptual deficiencies become audible.

Therefore, a *fast* simulation method that determines relevant outdoor propagation phenomena and delivers results that can be interpolated if simulation rates are insufficient is of elevated importance for real-time auralisation.

3.2 Wave-based simulation

The behaviour of sound is described by *wave mechanics*. Sound transmission can therefore be evaluated mathematically solving the *wave equation*, which includes partial differential terms and requires idealised environmental conditions. This topic is discussed in countless books on acoustics, for example, in the Encyclopedia on Acoustics by many authors [Rai98].

3.2.1 Closed-form solution

Theoretically, the wave equation can be analytically solved by closed-form functions. For instance, the *general solution* describes travelling sound pressure waves in one dimension and is relatively easy to comprehend for sine and cosine excitation functions (vibrations).

These formulas represent the fundamental elements of sound propagation and become complex when applied to problems in three-dimensional space. The integration of boundaries like, for example, a tilted wall already significantly increases the equation complexity. This aspect makes the closed-form solutions unwieldy for complex urban environments, although it is highly relevant for the detailed investigation of acoustic behaviour regarding simplified cases, like a single corner, a noise barrier or street canyons [Hor09]. In the context of outdoor sound, it is conceivable to employ such approaches to well-aligned problems, imposing strict patterns in certain dimensions of the built environment, for example, a raster of buildings or a repetition of street canyons [GHK19; Men+18].

The feasibility is limited to a selected number of simple scenarios, making closed form solutions useful to provide a reference in comparative studies of rudimentary outdoor problems regarding methods, measurements and models of diffraction.

3.2.2 Numerical solution

In practice, it is more efficient to discretise dimensions like space, time and frequency in order to linearise sections of a propagation problem. Numerical acoustic approaches use this to their advantage at the expense of partially exhaustive resource requirements with high demands on both memory and computation time.

For outdoor sound simulation, numerous attempts to apply numerical approaches have been pursued in the past, using the Finite-Difference Time Domain *(FDTD)* method and variants, receiving a lot of attention as described, for example, by Van Renterghem, Salomons and Botteldooren [VSB06]. Numerical solutions in the frequency domain, such as the Boundary Element Method *(BEM)* or the Finite Element Method *(FEM)*, are valuable tools to investigate the behaviour of complex structural and air-borne sound propagation problems at the cost of an exponentially increasing complexity regarding the desired frequency. A major advantage of numerical approaches is the capability to simulate the physical behaviour of sound adequately in a volume of manageable size, which can include highly complex surfaces[16] and inhomogeneous or non-isotropic properties. For example, Van Renterghem et al. [Van+12] employed various methods to study the performance of vegetated noise barriers on the reduction of road traffic noise and Kamrath et al. used a fast-multipole BEM as reference for investigations on noise barrier shapes and more complex situations with buildings [Kam+18].

On the one hand, low-frequency sound waves are subject to wave effects, which aligns well with the integration of diffraction in urban environments in this register. On the other hand, the complexity class narrows down the feasible frequency

[16] concerning the wavelength of interest

range dramatically. Viable solutions for extended urban environments are not able to progress into the mid-to-high frequency range of the human auditory system, which stands in conflict with the objective to design and implement a comprehensive urban auralisation system.

In particular, the attempt to implement a comprehensive yet fast sound propagation simulation of a realistic urban scenario – with the capability to provide real-time update rates – seems out of reach. Even with state-of-the-art high-performance computing capacities, it appears likely that many calculations must be performed offline and/or limitations and restrictions of various dimensions are required to perform an auralisation with an interactive component. Therefore, numerical methods for outdoor sound auralisation purposes are disregarded given the argument of limited flexibility and infeasible complexity for spatially extended urban scenarios.

3.3 Geometrical simulation

Geometrical Acoustics (GA) is a generic term for an entire acoustic field that regards the propagation direction of a sound wave in a fluid medium and interprets it as a ray or a particle. The proximity to geometrical optics, which is based on the assumption that light travels along a straight-line path in a homogeneous medium and changes direction at boundaries of dissimilar type, is apparent. Furthermore, the amplitude and phase may be affected by geometrical spreading, medium absorption and during interaction with boundary surfaces.

One of the most appealing properties of GA is the fact that geometrical propagation paths are independent of the wavelength. Propagation pathfinder algorithms generate an abstract, interim result during sound propagation simulation, which can be used for the entire frequency range of interest. However, it must be noted that GA methods ignore wave effects for the most part and lead to invalid results where these effects become important, predominantly in the low frequency region. Another considerable advantage of GA is the inherent focus on the acoustic transmission aspects of a source-receiver connection. The approach only calculates the contribution of each sound source to the regarded receiver location. Consequently, GA is best suited for virtual acoustic environments that require the calculation of sound at selected locations, considering a limited number of sources (e.g., in contrast to noise maps).

A good overview is given by Savioja and Svensson in [SS15].

3.3.1 Discretisation and quantisation

Digital systems require discretisation and quantisation. In the context of acoustics it is required to discretise the spatial and the temporal/frequency domain, and quantise representing values. On modern computers, the issue of limited memory may play a major role for high-performance applications, like physics simulations, but are of no concern for resolution decisions in GA algorithms and high-quality audio processing.

Geometrical values that represent locations and mesh points generally use 32-bit floating-point precision, which provides a sufficiently accurate resolution even for extremely large environments. However, numerical issues must be addressed in geometrical algorithms where appropriate, for instance, when small and large values are combined or divisions have small denominators leading to large intermediate values. To a certain extent, a geometrical threshold ε_{geo} can be introduced to avoid issues given a marginal spatial discrepancy (e.g., less than a millimeter displacement error). Serving as a short example, two planes may be considered co-planar and merged to a single plane if the dot product of the plane normals are closer to ≈ 1 than the threshold ε_{geo}, in order to avoid calculating an almost infinitely small diffraction component.

3.3.2 Deterministic solution

A deterministic geometrical solution solves theoretical propagation paths along line segments of least distance via a set of boundaries under the assumption of section-wise isotropic and homogeneous medium conditions. The concept is based on Fermat's principle and Snell's law and covers specular reflection, diffraction and refraction problems. A key feature is the unambiguity of the result, hence the attribute *deterministic*. It provides unequivocal reproducibility, which makes deterministic solutions appealing for the research community, and is often applied in noise prediction standards. Without the integration of sophisticated acceleration methods, a deterministic algorithms scales *linearly* with source-receiver pairs and *exponentially* with the number of boundaries, in principle, making them only feasible for very low orders in complex environments – or for low-to-mid orders in simple cases.

One of the most established and frequently implemented algorithm is the Mirror Image Model *(MIM)* [AB79; Bor84]. It constructs a substitute image of the sound source, if the sound wave interacts with a polygon or planar surface and effectively implements a *specular reflection*. Analytically, the image source must be projected to the opposite side of the reflection plane, hence the *mirror* connotation. If applied in an iterative fashion, a point cloud of mirror images can be spanned via the set of surfaces, and the coherent summation of each image source contribution (under free-field conditions) results in the desired simulation of the acoustic reflection pattern at a receiver location. The validity of each image source must be provided retrospectively. On the one hand, it must be ensured that the ray between receiver and image under test intersects within the bounds of the corresponding polygon by a point-in-polygon test. On the other hand, it must be assured that the free-path segments do not intersect with other polygons.

Another class of deterministic methods are volumetric tracing algorithms, such as the widely known beam tracing approach [Fun+04]. Here, propagation paths are determined by a spherical spatial separation from the environment, forming pyramids originating from a given location (i.e., with a triangle, rectangle or cone[17] base). By scanning the viewport of each beam, objects within volumes are detected and the beam resolution is refined until a beam only hits (or contains) one reflecting surface (or diffracting edge). At this point, the tracing is continued using a redirected beam volume. The implementation approaches show a certain similarity to methods originating from computer graphics, for example, in the projection algorithms required by object detection (scanning) routines in the frustum [LCM07; Tay+09a].

In the context of outdoor auralisation, diffracted sound is expected to play a key role. An interesting extension to the MIM is to integrate diffracting edges as mirrored diffracting edges into the pathfinder routine for segmented propagation paths [ESV21]. To consider segment-wise canonical wedge-type diffraction constellations helps to deterministically find the most important propagation paths, for instance, those with few reflection and diffraction orders. Therefore, Section 3.6 assesses the relevance and feasibility of this approach for the real-time auralisation of dynamic urban environments.

[17] Potentially non-deterministic and ambiguous if beams overlap

3.3.3 Radiance / energy exchange method

The *acoustic radiance transfer method* [Sil+07] is based on the idea that finite surface elements passively receive and emit acoustic energy, which was initially described in the context of computer graphics as *radiosity* method and subsequently applied to acoustic problems later [HN06]. Thus, elements exchange energy independently from the source and receiver locations. Consequently, a-priori knowledge on the amount of energy, exchanged between elements, is generated. However, in the acoustic domain, special attention must be drawn to the fact that energy is *distributed over time* due to the speed of sound, which cannot be regarded as infinitely fast. The radiance method requires stationary geometry on the one hand but breaks down the simulation procedure significantly on the other hand as the transmission problem essentially becomes an iterative lookup routine that accumulates energy at a location over time.

The capability of the radiance method to handle diffuse reflections in the first place makes it applicable to the diffuse sound field component, for example, in room acoustic simulations. Nonetheless, Siltanen et al. embedded an angle-dependent reflection kernel function, making it possible to represent specular portions by application of the algorithm towards non-diffuse components [Sil+07], which is not limited to indoor scenarios. However, wave effects are still largely ignored by the method.

Since outdoor acoustics exhibits diffuse reflections, the radiosity method and extensions to geometrical models have been investigated and integrated into simulation methods in the past, for example, by Can et al. [CFP15], Kang [Kan02; Kan00], Maillard et al. [MJ13] and Heutschi [Heu09].

3.3.4 Stochastic solution

Stochastic simulation methods using the principle of GA are essentially represented by *ray tacing* solutions applied to room acoustics [KSS68]. In short, rays or particles are emitted from an origin location and are traced trough the environment. The penetration of one or many sensitive surface patches or volumes, mostly detection spheres, is observed during ray casting. Rays are subject to directional modification as they travel through space. For example, a ray hitting a wall may be redirected based on a reflection function associated with the boundary element, that is not necessarily specular and can also split rays into fractions to incorporate diffuseness [Vor11]. Conceptually, a ray tracer conducts a *Monte Carlo* experiment and requires *convergence* for a generally valid result.

One problematic component is the suitable definition of the counting area or volume, which implies the risk of decreasing spatial accuracy when expanding and decreasing ray sampling resolution when contracting. However, the determination for a feasible set of parameters that approximates a sufficient quality as fast as possible requires experience and can also be enforced at the cost of higher computational load by selecting an exaggerated number of casted rays [Vor11]. If the ray count is subject to temporal aggregation due to back-reflecting surfaces (i.e., reverberant rooms), the stochastic nature of diffuse sound supports the validity of the method with increasing RIR length.

The complexity class of stochastic ray tracing exhibits a linear relationship regarding the number of boundary elements (e.g., polygonal surfaces). This is a unique position feature in acoustic simulation algorithms to date. The particularly fast calculation of ray casting enables to calculate plausible room acoustic simulation results at interactive rates [Pel+14]. Various extensions and modifications have been investigated to improve the perceptual quality and simulation accuracy (see, e.g., [Vor11; PSV]). Of relevance to outdoor propagation problems is the approach of Stephenson et al. describing an interesting idea to integrate diffraction into a stochastic ray casting algorithm by re-interpreting Heisenberg's uncertainty relation [Ste96]. Being in good agreement, at least for relatively simple situations, the rays are bent with respect to their original direction based on a function of distance to a pre-defined diffracting edge [Jud+16; WAS17; WSS18]. However, the principle of convergence, as required by Monte Carlo methods, must be ensured for outdoor scenarios, otherwise the reliability of the simulation result remains unclear.

3.3.5 Hybrid methods

Some attention has been drawn to hybrid simulation methods to combine the advantages of two separate approaches and overcome inaccuracies regarding certain aspects of spatial, temporal or frequency resolution. For instance, a specular approach and a diffuse method can be combined to individually simulate the early part and the reverberant decay of RIRs, respectively [Nay92; SV07; SP09]. It is also conceivable to process different frequency ranges via separate algorithms and subsequently blend the individual results [AV14a; AV14b].

In outdoor acoustics, it is appealing to simulate the low frequency range by numerical methods and the high frequency range by GA methods. In this way, wave effects can be included where most relevant, while maintaining a broad-band auralisation. However, the computational complexity of a hybrid outdoor simulation approach evidently increases as two routines must be applied [Hor16]. This issue is particularly relevant for real-time auralisation applications. Nonetheless, a complementary algorithm that enhances deterministic solutions that calculate specular reflections and diffraction by a method providing diffuse sound field components is deemed more suitable for outdoor scenarios.

3.4 Outdoor propagation models

The class of GA algorithms, as introduced in Section 3.3, provides geometrical representations of sound paths, which have to be interpreted by physical propagation models. These models describe the alteration of acoustic waves by mathematical functions and are valid for a stationary scenario. In the context of dynamic auralisation, quasi-stationary *snapshots* are often assumed and blending of subsequent simulations is applied to trigger scene changes over time (cf. Section 4.3.1).

During propagation, different wave interaction types with the environment are encountered – and the majority of them attenuates sound frequency-dependently. In the scope of auralisation, the necessity to integrate spectral shaping has an unwanted side effect of requiring a convolution function in order to incorporate the modifications. Filter units that effectively apply frequency shaping, some of them disregarding the phase, are computationally costly and consume resources in the audio context where efficiency is imperative (cf. Figure 3.1). However, because frequency-dependent attenuation plays a major role in outdoor sound, it is essential to include this behaviour in an auralisation approach.

Among all the propagation phenomena, special attention is drawn to models of acoustic diffraction at wedge-like edges. It is anticipated, that an auralisation application for outdoor environments requires a comprehensive handling of sound that comes from partly or fully occluded sound sources. A profound concept to realise this effect with respect to quality and calculation efficiency is decisive, assuming that fast moving sound sources appear and disappear over the course of time.

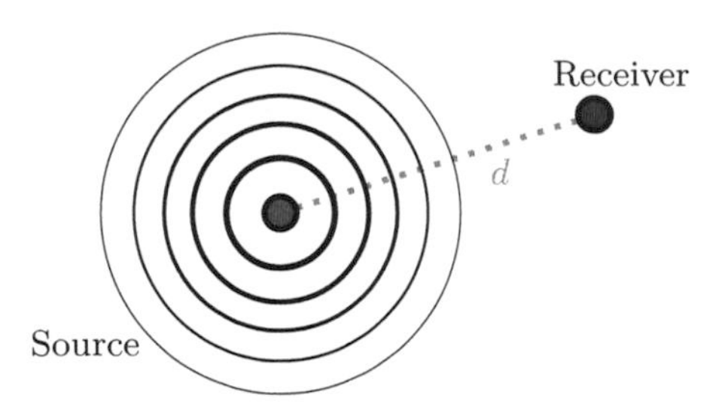

Figure 3.2: Illustration of spherical spreading loss.

Relevant acoustic phenomena occurring in outdoor environments are briefly explained in the following sections. If not otherwise stated, attenuation coefficients are determined in the frequency domain for the reason that subsequent alterations during propagation can be efficiently combined by component-wise (complex-valued) multiplication. For more details on fundamental auralisation topics, the reader is referred to the literature (e.g., [Vor11; Beg00]).

3.4.1 Spreading loss

Spreading loss is a general wave effect that describes the energy decrease observed with increasing distance and depends on the wave type, as illustrated in Figure 3.2 for a spherical wave emitted by a point source. Applied to longitudinal spherical sonic waves under isolated homogeneous conditions, the sound pressure fluctuation amplitude $\hat{p}$ diverges reciprocally over distance d, yielding

$$\hat{p} \propto \frac{1}{d} \,.$$

Hence, to make a virtual distance audible in an auralisation algorithm, the emitted sound pressure signal must be divided by the scalar value of the Euclidean length of a vector between the sound source and the observation point. The spreading loss is not depending on the content of the signal.

3.4.2 Medium absorption

The propagation medium absorbs energy from the sound pressure wave depending on the *frequency* content of the signal. Outdoors, a non-adiabatic behaviour of

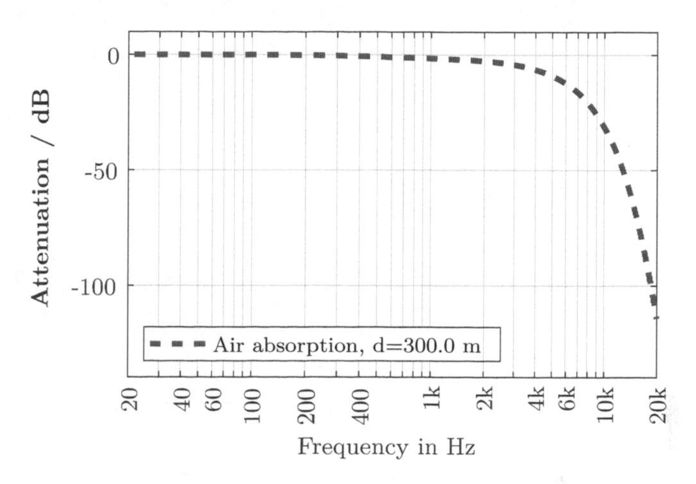

Figure 3.3: Attenuation transfer function of air at 20 °C and 80 % rh at a distance of 300 m.

air at a given temperature, moisture and static pressure is assumed, which can be approximated by a frequency-dependent real-valued attenuation factor

$$a_{\text{abs}}(f) = d \cdot \alpha_{\text{medium}}(f, T, rh, P) \ .$$

specified by the frequency f, the temperature T, the relative humidity rh and the static atmospheric pressure P specified in units of dB/m [KB20]. Figure 3.3 depicts the attenuation transfer function of air at 20 °C, 80 % relative humidity and standard atmosphere at a distance of 300 m according to ISO 9613 [ISO93; ISO96].

3.4.3 Specular reflections

If a sound wave encounters a surface, for example, a building facade or a soft grassy ground plane, the impedance discontinuity causes a reflection that is redirecting the wave. This results in a wave with an equal or lower amplitude and a phase shift depending on the frequency. For environmental acoustics in an urban surrounding, precisely determining the reflected sound energy is important as hard surfaces made of stone and concrete are almost non-absorbing, see Figure 3.4. In the context of outdoor simulation using GA algorithms, the absorption factor

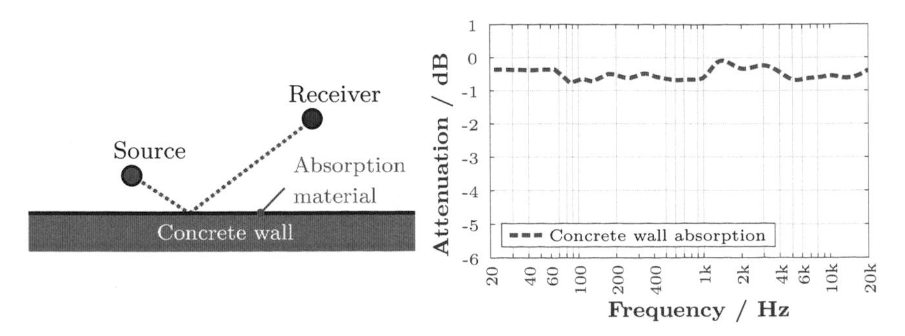

Figure 3.4: Specular reflection of a sound wave that is attenuated by absorption of the material.

is not necessarily evaluated in terms of (angle-dependent) impedance functions, but is provided as a look-up table of coefficients with a given frequency resolution, for example, energetic third-octave band spectra, assuming a perpendicular angle of incidence. These spectra are either approximated by models or have been established by measurements and stored in a database. During the simulation, all reflections at surface polygons are tracked and linked with the corresponding surface material. In the propagation modelling stage, these assignments are required to query the material database and integrate, for example, absorption spectra.

3.4.4 Diffraction

Any non-planar junction of surfaces, like an edge or a corner, corresponds to a disturbance of the incident sound wave that differs from a pure specular reflection. The underlying physical paradigm is called *diffraction* and is defined as an additional frequency-dependent sound field component with attenuated amplitude and a phase shift. Diffraction solutions exist for canonical cases such as screens and wedge-shape objects of finite or infinite length. Geometrically interpreted, a diffraction problem can be separated into four regions, in which different sound field components must be combined. These regions are determined by the source location, as indicated in Figure 3.5, and leads to the corresponding transfer function, exhibiting a low-pass character modelled after three established methods (cf. Section 2.3). In the context of urban sound, diffraction contributions are present in every region, except for solid structures. Depending on the receiver location, these different components must be addressed by a model. In the literature, and

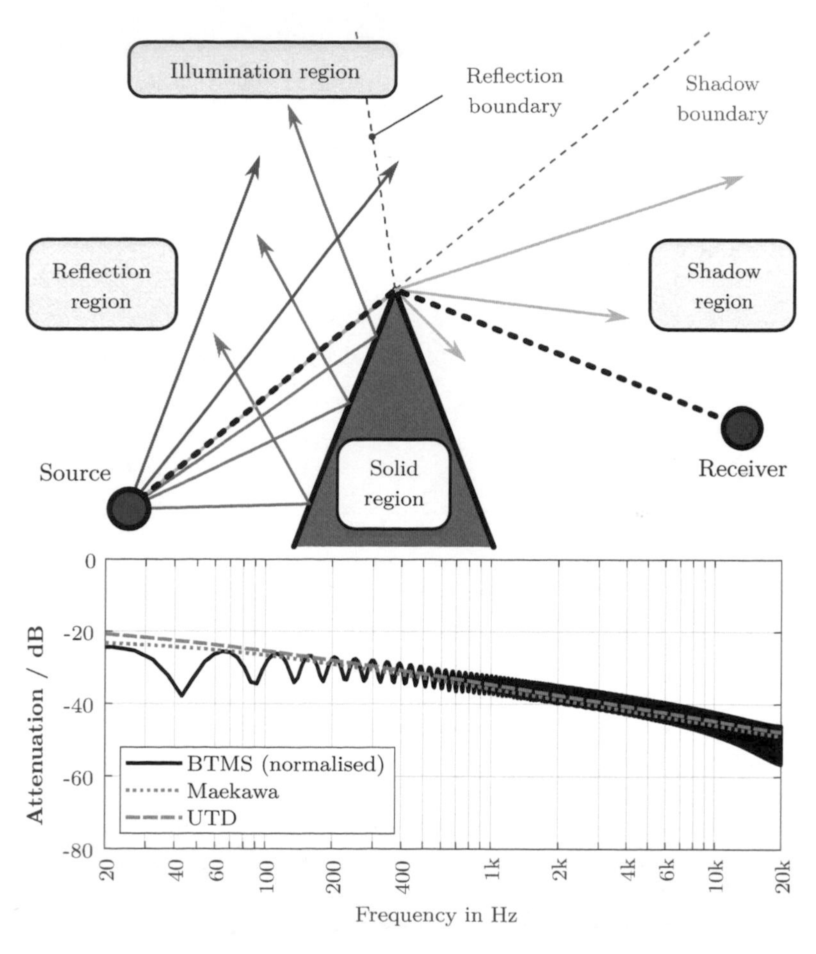

Figure 3.5: Simple wedge-type edge diffraction scenario with corresponding attenuation transfer functions of three different models.

in particular in the context of noise assessment, diffraction problems are often reduced to situations where the receiver is located in the *shadowed area*. This approximation is comprehensible insofar as that neither direct sound nor reflected energy is present, making the diffracted sound the only contributor to the total sound field in these regions. However, it is reasonable to include all components for auralisation purposes to maintain continuous sound field properties, as discussed in Section 3.5. Also, acoustic energy redirected backwards adds a diffusion

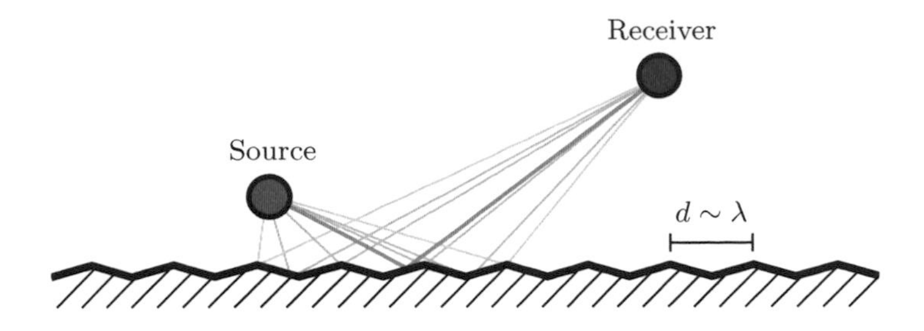

Figure 3.6: Illustration of a diffusely reflected wave reaching the receiver via a rough surface, with similar dimension in wavelength λ and spacing.

component, referred to as *(back-)scattering*, which is especially prominent for rough surfaces depending on the wavelength.

3.4.5 Diffuse reflections

Sound waves hitting rough surfaces are scattered in a complicated way, if the corrugation dimensions are of comparable magnitude to the corresponding wavelength. In contrast to specular reflections, which can be precisely determined analytically, the scattered energy is treated by means of stochastic processes. A probability functions is evaluated with respect to the likelihood of scattering sound waves in a certain direction, as per Lambert's law. Analogous to absorption spectra, surfaces can be assigned random-incidence scattering coefficients obtained from simulations or measurements, which can be accounted for by propagation modelling algorithms.

3.4.6 Directivity

In virtual acoustics, the *direction-dependent* radiation function of a sound source, that is, it's *directivity*, characterises radially diverging sound emission. Directivities are often represented as frequency-dependent but distance-*in*dependent directional datasets that only satisfy far-field conditions. However, many different representations are conceivable and directivities are mentioned here as an example. Although they require attention in the propagation modelling routine, no further generalisation appears appropriate at this stage as the actual imple-

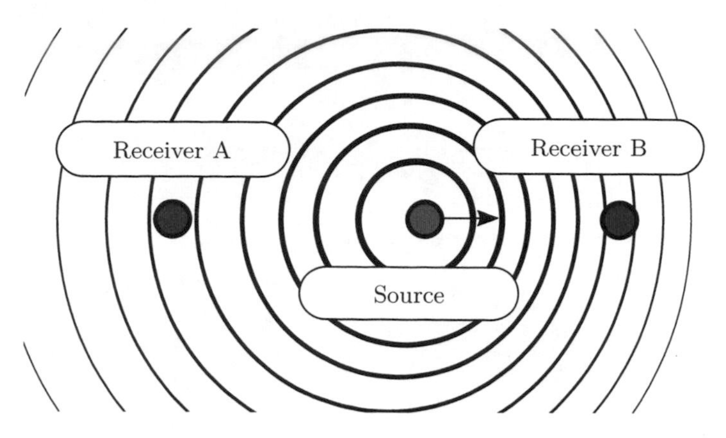

Figure 3.7: Illustration of the Doppler shift.

mentation is entirely context-dependent and the format of available data usually dictates the designated method.

3.4.7 Doppler shift

The Doppler shift is a *time-variant* acoustic effect that is observed given relative movement against a propagation medium with finite propagation speed. It requires the consideration of motion over time below the speed of propagation, as illustrated in Figure 3.7. If a source moves as indicated, emitted sound pressure waves are squeezed towards the direction of motion and stretched in the opposite direction. In this scenario, receiver A and B experience sound waves with lower and higher wave lengths, respectively, resulting in lower and higher frequencies/pitches. The same phenomenon occurs, if a receiver moves relative to the propagation medium. Physically, the combined result leads to a complicated, temporal relation between source motion from the *past* – dating back to a point in time that corresponds to the elapsed propagation delay – and the *current* receiver motion. Although it appears straight-forward to implement the phenomenon by joining both motion vectors at the corresponding time and use the resulting, *relative* motion to scale the frequencies according to Doppler, an error is introduced. This approach neglects the fact that the emitted source signal is attenuated by frequency-dependent processes subject to source-related Doppler only. Theoretically, the receiver-induced time-variant Doppler shift must be applied after propagation modelling as it is not valid to treat the situations

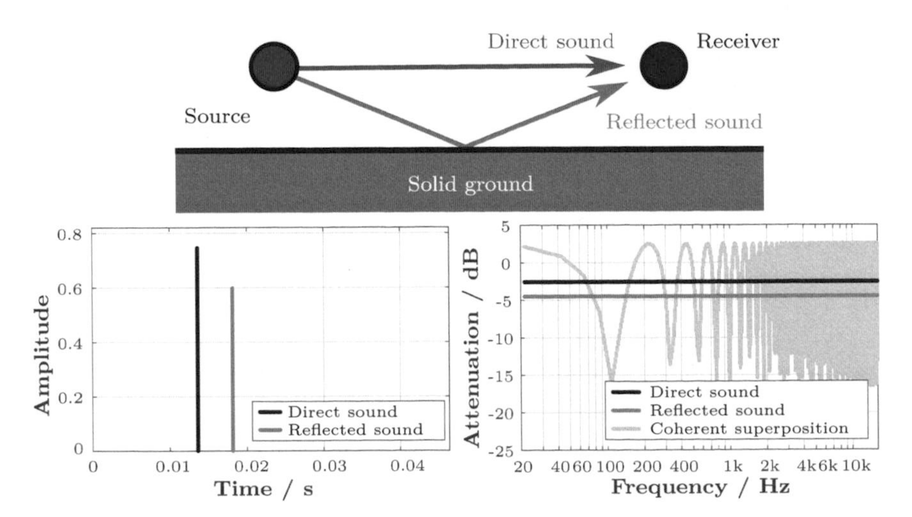

Figure 3.8: The coherent superposition of a direct and reflected sound path contains two impulses in the IR (left) and results in a comb-filter attenuation pattern, as depicted in the *TF* (right).

as a Linear Time-Invariant *(LTI)* system. A comprehensive time-variant solution to correctly combine source and receiver Doppler effects considers a *scaled* propagation spectrum depending on the source-medium motion, see Section 4.3.2.

3.4.8 Coherent superposition

The superposition of sound pressure waves results in a complicated, frequency-dependent interference. In outdoor scenarios, signals from *different* sound sources can be considered as incoherent, and energetic addition applies (cf. Section 2.5). However, signals from the *same* sound source reaching the receiver via different paths exhibit a strong resemblance and require coherent superposition. To illustrate this property, a common outdoor situation where sound reaches a receiver via a direct path and a reflected path over ground is sketched in Figure 3.8. While the IR representation is relatively easy to comprehend, the effects of adding two similar, slightly time-delayed sound pressure signals exhibit a highly fluctuating spectrum (comb-filter structure). The result reveals constructive and destructive interference depending on the frequency (or the wave length) and the path lengths difference. Since the human perception is susceptible to frequency-selective sound reinforcement and cancellation, a comprehensive auralisation application requires models considering propagation paths consistently and superposing the individual contributions to the overall sound field at the receiver coherently. A comparison of a measured and simulated scenario accounting for these aspects is provided in Section 5.3.1.

3.5 Diffraction models for auralisation

In outdoor sound auralisation under real-time constraints, an appropriate *diffraction* handling is crucial. A simple yet accurate model is needed to provide a plausible rendering with fast adaption capability. From the introduced diffraction models in Section 3.5, the Uniform Theory of Diffraction (UTD) method is deemed most relevant (cf. Section 2.3.1), mainly due to its core benefit to maintain sound field *continuity* in critical regions. To emphasise the relevance of these features, a more elaborated investigation in the context of auralisation is necessary. So far, the direct sound discontinuity and reflection discontinuity of canonical scenes as shown in Figure 2.4 have been proven to be accurately compensated (see also plot in Figure 2.5).

To broaden the applicability of the UTD method for arbitrary polyhedral environments including multiple reflections and diffractions, the scenario of Figure 3.9 is considered. The situation is comparable to that of Figure 2.4, however, a ground plane is included and the roles of source and receiver are switched to represent a more realistic urban situation.

In this scenario, the trajectory follows an arc around the edge (two-dimensional cross-section) with regions of different sound field combinations (labels I-IV). Without the ground plane (face 3), the results are comparable to those of Figure 2.5, where the contribution of the image receiver R_2 is already accounted for (R_2 is independent of the ground plane). However, the modified situation now includes images of the receiver with respect to the facade (face 2) and ground plane (face 3), indicated by the circles labelled R_3 and R_{23} / R_{23} (coincident images). With the source's transition from section II to section III, the ground reflection immediately becomes audible (direct line-of-sight to image receiver R_3). This discontinuity is *not* covered by the diffraction component of the wedge, since the canonical situation is assembled by main face 1 and opposite face 2 sharing the edge with apex point M – and is not aware of the ground plane (face 3). Constructing an image basically removes the corresponding mirroring primitive. Applied to the current situation, R_3 removes the ground plane, effectively generat-

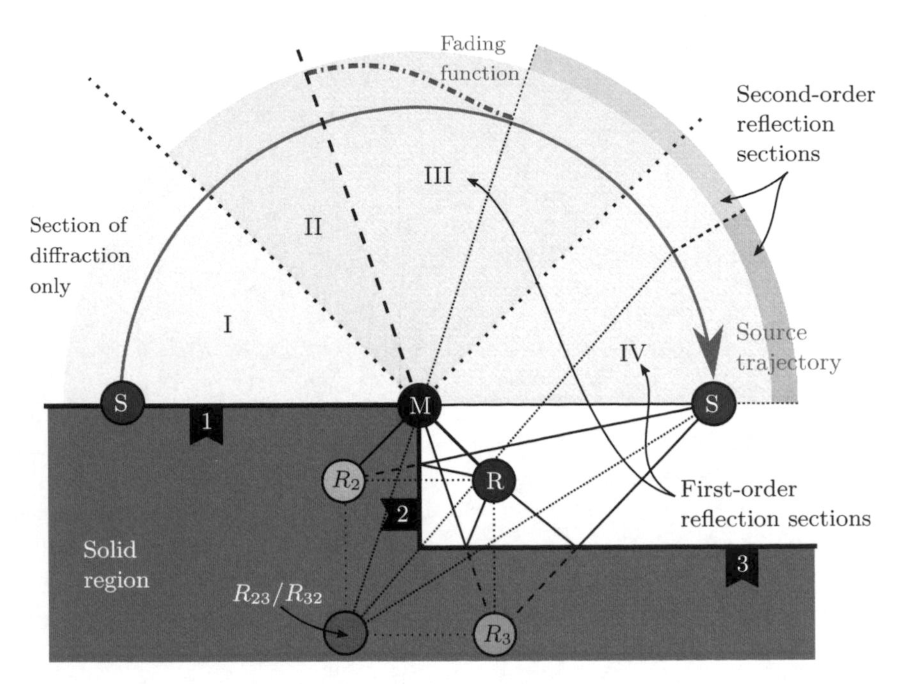

Figure 3.9: Example scene with a ground reflection, showing sections of different sound field components indicated along the trajectory of source S. The corresponding acoustic modelling is depicted in Figure 3.10.

ing a new problem that can be expressed by a canonical edge diffraction scenario, defined by the initial wedge (face 1 and 2) and the image receiver R_3. Hence it is concluded, that the diffraction path from source S via apex M to the image receiver is responsible for a smooth transfer at this critical boundary from II to III, making it imperative to include the diffraction component with the image receiver involved. Regarding the next transition from region III to region IV, introducing a direct line-of-sight to the image receiver R_2 (generated by face 2), the diffraction component of S via M to R maintains continuity (analogue to the situation in Figure 2.4), as this face is part of the wedge. However, considering that the ground plane and the facade form a right-angle corner, second-order reflections are expected as soon as the source appears in the sections that are indicated by green bands. Conveniently, the diffraction calculation with the image receiver R_3 automatically integrates the second-order reflection R_{32} and includes the corresponding diffraction compensation term at the reflection boundary (transition inside region IV). While this appears desirable at first glance, the extension

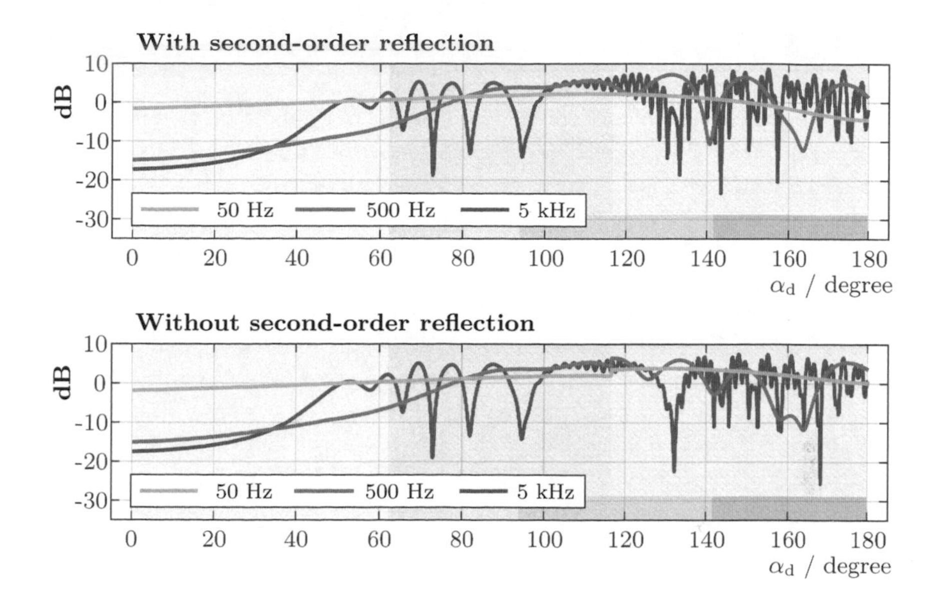

Figure 3.10: Transfer functions over trajectory angles for the diffraction problem with ground reflection shown in Fig. 3.9 (representative frequencies), with (top) and without (bottom) contribution of the second-order reflection.

with a first-order receiver image now implies, that a second-order reflection must be resolved. In other words, the consideration of a first order reflection leads to the requirement of a second order reflection to satisfy the sound field continuity principle. That said, such a recursive relation is also present in the canonical wedge problem, as the *diffracted* sound field requires the provision of the *reflected* sound field component at the reflection boundary transition. This leads to the conclusion that any inconsiderate limitation of the reflection and/or diffraction order inevitably results in a discontinuity.

A solution to this problem is proposed by *locking* the number of reflections and diffractions to a fixed maximum order, while only allowing one further diffraction of the combined order. In the modelling, if a maximum-order propagation path contains a diffraction as the last segment, continuity is ensured by a *fade-out function* suppressing the increasing field component towards the reflection boundary that cannot be provided.

To deliver a smooth fade-out, a scalar function is suggested to be multiplied with the complex-valued UTD coefficients in the section between the shadow boundary and the centre of the section between shadow boundary and reflection boundary. Considering Figure 3.9, a locking to the first order must be used, which accepts a single reflection or a single diffraction, and one subsequent diffraction that is partly faded to suppress further compensation of second-order reflections that are not available. In this case, the fading function is applied in the separately regarded diffraction problem involving the source and the first-order reflection represented by the substitute receiver R_3. The single-sided window begins with the plateau at the left-hand beginning of sector III (light-purple area), and fades out approximately towards the middle (dotted line), where it fully suppresses further diffraction coefficients. The fading function is indicated as purple dash-dotted curve labelled correspondingly. The dotted line inside the light-purple section III represents the reflection boundary of the image receiver R_3. Theoretically, it corresponds to the location where the section with the missing second-order reflection begins, as indicated by the light-green band in the top right corner.

To visualise the discontinuity problem in the acoustic modelling, Figure 3.10 depicts the complete, smooth series (top sub-plot) and the truncated, discontinued curve (bottom sub-plot), showing transfer function magnitudes over the source curvature for representative frequencies of 50 Hz, 500 Hz and 5 kHz. Compared to the single-wedge problem, particularly inside the reflection area ($\alpha_\mathrm{d} > 63°$) typical interference pattern caused by superposition of coherent transfer paths are found. The first plot shows the complete solution including the second-order reflection sound field component, which must be manually added to the transfer function after passing the image reflection boundary at an angle of approximately $\alpha_\mathrm{d} = 116°$. It can be clearly seen, that the superposition of the source-receiver and source-image-receiver paths with corresponding reflection contribution results in a continuous progression of the three frequencies. However, comparing the second plot where the next-order reflection is *not* considered, since it is not available because it exceeds the pre-defined maximum order of reflections, an undesired discontinuity at the respective boundary transition area occurs. Although the step size appears small, the magnitude spectra change by several decibels, which is perceived as a jump in the loudness of the auralisation result.

A solution to intercept discontinuous diffraction coefficients of the UTD when approaching the reflection zone is proposed below. Figure 3.11 depicts the modification of the diffraction coefficients by separately investigating the contribution of the first-order reflection, for example, by the additional path from source to image receiver R_3. The first plot shows the individual target diffraction problem

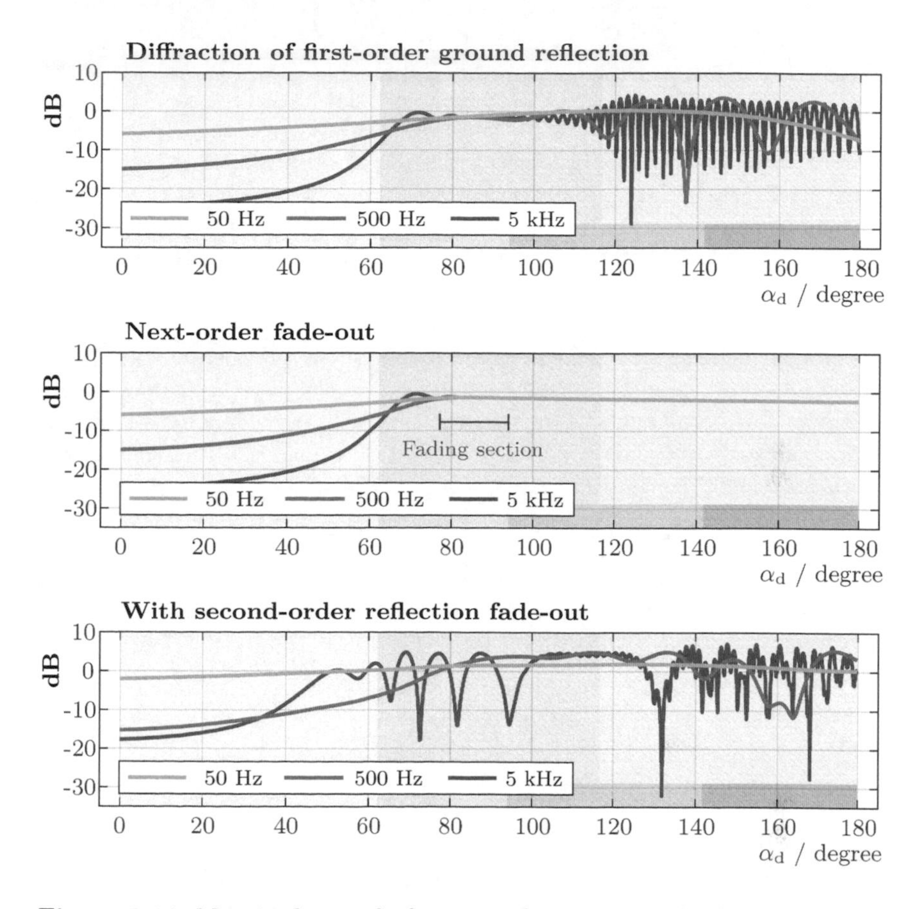

Figure 3.11: Magnitude transfer functions of trajectory angles for the diffraction problem with ground reflection after Fig. 3.9 (example frequencies) with original diffraction component (top), proposed fade-out towards next-order reflection boundary (centre) and resulting continuous sound field (bottom).

including the ground image, which requires the second-order reflection contribution for continuity. The second plot shows the diffraction component with the fading applied, which becomes active when reaching the corresponding illumination area at approximately $\alpha_d = 64°$. A cosine-square function blends out and suppresses values half-way towards the reflection boundary at $\alpha_d \approx 88°$, at which point the direct sound component is the only remainder. The reason to choose this particular fading range is that the UTD coefficients are vanishingly small, yet start to increase around this α_d. Additionally, the evaluation of the centre between shadow and reflection boundary is easily determined by geometrical considerations. The final Transfer Function *(TF)* for the representative frequencies is depicted in the third plot of Figure 3.11. In comparison to the second plot of Figure 3.10, the discontinuity is removed by the coefficient fading function in the critical section without affecting the curve progressions. Consequently, the fading function must be applied if a maximum-order image is involved to address the missing contribution of the next-higher image order.

The fading approach discussed by Tsingos et al. [Tsi+01] completely neglects the diffraction component in the illuminated area and permits a significant error at the shadow boundary. In contrast, the proposed procedure maintains the accurate acoustic attenuation values in the vicinity of the shadow boundary and only fades out transfer function values where a missing sound field component is not available in the first place due to a limited image order.

These findings underline that the comprehensive integration of acoustic diffraction based on the UTD model requires a deterministic pathfinder algorithm that can deliver the entirety of propagation paths, including arbitrary combinations of reflections and diffractions.

3.6 Image-edge pathfinder algorithm

In the course of this thesis – and in prospect that a deterministic propagation path algorithm combining reflections and diffractions is necessary – a *pathfinder* has been implemented and published as *Image Edge Model* (IEM) [ESV21]. The method finds the shortest geometrical path via scene primitives of a polygonal mesh with reflecting surfaces and diffracting edges, bearing Fermat's principle in mind. Initially, equality of incidence and exit angle of a wavefront normal is considered. Transferred to diffraction problems, the redirected wavefront must fulfil Keller's law of diffraction in sufficient approximation with respect ot the edge direction [Kel62]. This assumption is comparable to the incoming and outgoing angle constraint for the wavefront normal with respect to the face normal found in specular reflections fulfilling Snell's law [AB79]. In fact, the proposed approach can be seen as an extension to the classical MIM by Allen and Berkley, adding mirror edges to the equation. The idea was initially mentioned by Tsingos et al. and has been integrated into a beam-tracing approach [Tsi+01]. Funkhouser et al. indicated as well, that it is possible to incorporate specular reflections between

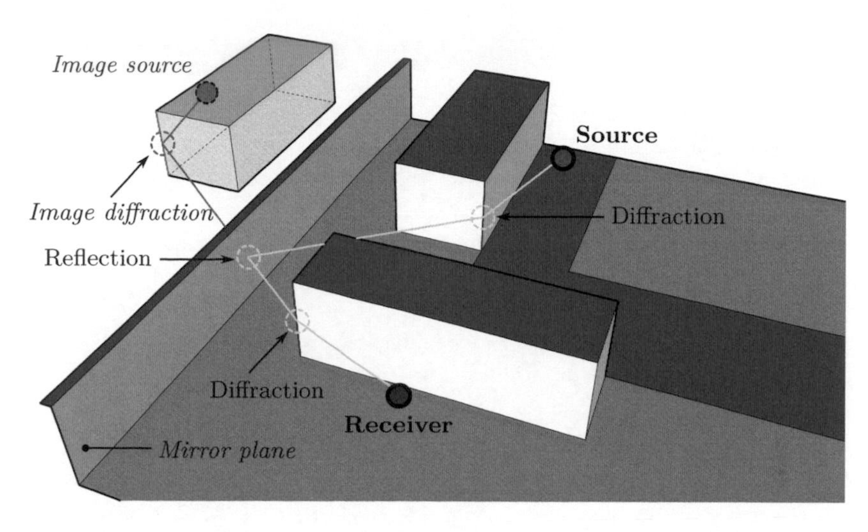

Figure 3.12: A pathfinder determines propagation paths, consisting of arbitrary sequences of reflections and diffractions. In the presented scenario, a diffraction-reflection-diffraction path is shown. A fast solution can be found by applying a mirroring operation at the reflection plane.

two diffractions in [Fun+04]. Mathematically, the imposed angle requirements can be described by a set of linear equations – a problem that can be solved by matrix formulation and inversion. The proposed novel pathfinder adds a new way of solving this Equation System of Equal Angles *(ESEA)* by rearranging the matrix to a set of segment-wise independent equations in order to avoid the otherwise computationally expensive solution by a matrix inversion.

The implementation of a pathfinder following the IEM approach is motivated by situations that require to treat reflections and diffractions separately to determine important propagation paths. Figure 3.12 depicts such a scenario and demonstrates a propagation path that can only be found with an approach that integrates mirror sources and mirror edges. In the absence of simpler paths including, for example, a single reflection followed by a single diffraction, these complex paths become highly relevant. This bears in mind the sound field continuity requirement of dynamic auralisations in order to provide acoustic energy in areas where, in the worst case, the sound source becomes abruptly inaudible [Ant+12] (see also Section 5.4). Theoretically, it is conceivable that the majority of potentially important propagation paths include very few reflections and one or two diffraction components. Many of these paths can be found by a subdivision approach that deals with reflections in the vicinity of the source and the receiver first, and subsequently identifies diffractions among the images. In this way, it is possible to find propagation paths with reflections at the beginning and at the end, and diffractions in-between. Although the subdivision approach can potentially produce a sufficient path list, it is unsatisfactory that the method lacks to find all propagation paths, for example, those consisting of a reflection *between two diffractions*. Such a scenario can be easily constructed, as shown in Figure 3.12. This is especially true if it is desired to auralise a sound source *in motion*, since the feature to deterministically find arbitrary reflection-diffraction propagation paths is key to auralise a *continuous sound field* without areas of total silence and abrupt loudness jumps.

For this reason, the implementation of a comprehensive pathfinder algorithm has been given priority in the context of outdoor sound auralisation.

Although dynamic scenarios were the focus of the development, the proposed procedure based on the IEM requires a-priori mesh information to accelerate the propagation paths determination from an origin to a target location. Hence, the virtual environment has a static geometry, and dynamics are limited to sound source and receiver motions. To a certain extent, it appears obvious that the urban built environment is regarded as static. However, this simplification hinders to include interactive geometry modifications and therefore discards the immediate (acoustic) observation of an intervention. For example, it is not possible to realise a VR application that allows to place a noise barrier and adjust its height while listening to the result at the same time. Although this limitation is undesirable it was a necessary compromise to approach real-time capability of the IEM procedure. However, an approximate alternative solution to tackle this problem will be presented in Section 3.7.

It was anticipated that the pathfinder algorithm delivers a vast amount of theoretical propagation paths, considering the inherited exponential growth rate as a function of reflection and diffraction orders. Therefore, technically and perceptually motivated abortion methods were developed to intervene as early as possible during path finding, favouring minimal computational efforts. This idea also challenges classical termination criteria like image-order capping. From a technical viewpoint, it may seem reasonable to stop a pathfinder routine after several reflections and diffractions, but it remains a questionable method in terms of perceptual aspects. The risk of ignoring a perceptually relevant propagation path that can only be found after several reflections and one or more diffractions ought to be minimised.

3.6.1 Fundamental principle

The classical MIM projects a sound source perpendicular to the reflection plane of a surface, represented by an arbitrary polygon, to the other side of the plane at the point of same distance to the plane [AB79; Bor84]. This new image is a substitute that generates an equivalent acoustic field in front of the plane and within the spatial volume that is in direct line-of-sight from the image location through the polygon. The location where an imaginary ray between image source and sound receiver intersects the plane is the reflection point and corresponds to the shortest path from the source to the receiver via the intervening reflecting plane. As long as the point lies within the boundaries of the polygon and does not intersect with other objects, the path is valid – a process that is referred to as *visibility test*. This procedure can now be continued with this image source via another polygon to consider a second reflection, and repeated for all reflecting surfaces up to the desired order. Without limitation (except for a reversed direction), the MIM and the proposed IEM produce equivalent propagation path results, if the roles of sound sources and receivers are switched. For reasons of simplification, it is assumed that a propagation path origin corresponds to the sound source for the descriptions below.

From a more generalised perspective, an imaging approach applies a mathematical reflection, since it maps the Euclidean space in front of the reflection plane to the back-side, while making the initial plane redundant. This kind of transformation preserves the Euclidean distances as well as relative angles (global isometry), but changes the orientations, as depicted in Figure 3.13. Since the transformation is bijective, a function inverse exists and relations of the figure (mirrored object) apply likewise in the original space.

The solution to a geometrical propagation path problem is initiated from the *(multi-)mirrored space*, while the final locations (points in space) are determined by *unfolding* the mirror-space parameters back towards the original space for each order in a reverse manner (inverse transformation). The principle of a congruent transformation encourages to apply the mirroring idea not only to the source and its images but basically to any constellation including edges subject to diffraction [Tsi+01]. The procedure replaces every specular reflection

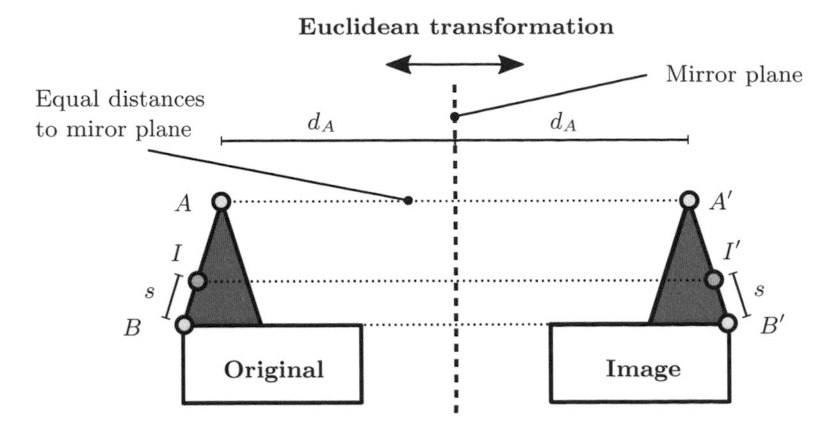

Figure 3.13: Schematic depiction of a Euclidean transformation applied to the original object, resulting in a mirrored image with preserved distances and relative angles, but changed orientation.

of the original situation with a mirrored space, until all reflections have been taken into account. The resulting representation is a repeatedly folded space that may be difficult to grasp. However, it purely consists of canonical diffraction problems that can mathematically be solved by an equation system formulating the sequential angle conditions for all involved edges in the corresponding space. The solution of the equation system delivers the parameters which are required to reconstruct the actual geometrical path in the original space. If the reconstructed interaction points of the geometrical path (sequence of reflection points and diffraction points) are within the bounds of the corresponding shape (polygon or edge) the problem is solved and determines the geometrical path of least Euclidean distance under the assumption of equal angles for incidence and exit wavefronts, as discussed in Section 3.6.2. Finally, it is necessary to validate that no path segment is interfering with other objects, requiring to perform a visibility test.

3.6.2 The equation system of equal angles (general ESEA)

Mathematically, the ESEA describes the constraints imposed on the incoming and outgoing wavefront normals regarding the respective primitives in the sequence of mesh items of a propagation path candidate [Tsi+01]. Under the assumption

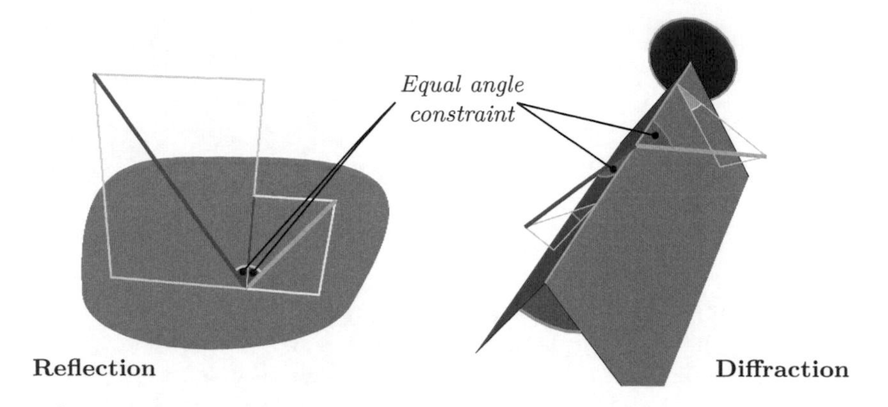

Figure 3.14: Canonical reflection and diffraction scenarios with indicated angles that must be equal according to the ESEA.

of far-field conditions, the solution delivers the shortest path via the primitives considering the Euclidean distance metric, as demanded by Fermat's principle.

The intersection point representing a specular reflection is referred to as *reflection point*. The point where a diffracted path crosses an edge is referred to as *apex point*. To generalise these special locations, the corresponding points will be called *interaction points* in the following. Every geometrical propagation path consists of a sequence of at least one source location at the beginning and one receiver location at the end, with no or any number of interaction points in between, each representing either a reflection or a diffraction.

To derive the system of equations, the initial requirements of a canonical reflection and a diffraction problem is considered, as shown in Figure 3.14. Locating the interaction point I on a reflecting plane F with the normal unit vector $\overrightarrow{n}$ connecting a point of origin A with a target point B in the initial Euclidean space specifies the angle constraint, formulated as vector products, yielding

$$\frac{\overrightarrow{IA}}{\|\overrightarrow{IA}\|_2} \cdot \overrightarrow{n} = \frac{\overrightarrow{BI}}{\|\overrightarrow{BI}\|_2} \cdot (-\overrightarrow{n}), \tag{3.1}$$

with $\|\cdot\|_2$ representing the L^2-norm. Accordingly, locating the interaction point I on a diffracting edge E with the edge unit direction vector $\overrightarrow{e}$ yields

$$\frac{\overrightarrow{IA}}{||\overrightarrow{IA}||_2} \cdot \overrightarrow{e} = \frac{\overrightarrow{BI}}{||\overrightarrow{BI}||_2} \cdot (+\overrightarrow{e})\,. \tag{3.2}$$

If A and B are locations of source and receiver, the determination of the only unknown variable, the interaction point I, can be found by geometrical considerations (see, e.g., Appendix in [Tsi+01]).

A slightly more complex situation already requires an analytic routine, for example, with two diffractions at linearly independent edges, as two equations with inter-depending variables must be solved. A general formulation corresponding to a sequence of $k \in K, K > 0$ interaction points I_k subject to reflection or diffraction with the respective unit direction vector $\overrightarrow{m_k}$, connecting a source location S and a receiver location R yields

$$\begin{aligned}
\frac{\overrightarrow{I_1 S}}{||\overrightarrow{I_1 S}||_2} \cdot \overrightarrow{m_1} &= \frac{\overrightarrow{I_2 I_1}}{||\overrightarrow{I_2 I_1}||_2} \cdot (\pm\overrightarrow{m_1}) \\
\frac{\overrightarrow{I_2 I_1}}{||\overrightarrow{I_2 I_1}||_2} \cdot \overrightarrow{m_2} &= \frac{\overrightarrow{I_3 I_2}}{||\overrightarrow{I_3 I_2}||_2} \cdot (\pm\overrightarrow{m_2}) \\
&\vdots \\
\frac{\overrightarrow{I_{K-1} I_K}}{||\overrightarrow{I_{K-1} I_K}||_2} \cdot \overrightarrow{m_K} &= \frac{\overrightarrow{I_K R}}{||\overrightarrow{I_K R}||_2} \cdot (\pm\overrightarrow{m_k})\,.
\end{aligned} \tag{3.3}$$

The decision for the sign of direction vectors $\pm\overrightarrow{m_i}$ depends on the type of i-th interaction. The vector notation $\overrightarrow{I_k I_{k-1}}$ describes the *directions* from the interaction point I_{k-1} to I_k of a path segment. The same applies for $\overrightarrow{I_1 S}$ and $\overrightarrow{I_K R}$ with the exception that one interaction point is substituted by S, the source location, and R, the receiver location. The ESEA can be formulated as a matrix product with the solution by its left inverse, yielding

$$A \cdot x = d \quad \Rightarrow \quad x = A^{-1} \cdot d\,. \tag{3.4}$$

The equation system forms a symmetric, tri-diagonal matrix, because the k-th interaction point term locally depends on the previous and succeeding value:

$$A_K = \begin{pmatrix} a_0 & b_0 & 0 & \cdots & & & \cdots & 0 \\ b_0 & a_1 & b_1 & 0 & \cdots & & & \vdots \\ 0 & b_1 & a_2 & b_2 & 0 & \cdots & & \\ \vdots & 0 & b_2 & a_3 & b_3 & 0 & & \\ & & \ddots & \ddots & \ddots & \ddots & \ddots & \vdots \\ & & & 0 & b_{K-2} & a_{K-3} & b_{K-3} & 0 \\ \vdots & & & & 0 & b_{K-3} & a_{K-2} & b_{K-2} \\ 0 & \cdots & & & \cdots & 0 & b_{K-2} & a_{K-1} \end{pmatrix} . \tag{3.5}$$

The equation system can be solved by Gaussian elimination, however, a simplified routine applies [CD17]. To accelerate the calculation, a re-formulation of the symmetric, tri-diagonal matrix is required to form coefficients following

$$a'_k = 1 \qquad \text{for} \quad k = 0, 1, ...K - 1 \tag{3.6}$$

$$b'_k = \begin{cases} \frac{b_0}{a_0} , & \text{for} \quad k = 0 \\ \frac{b_k}{a_k - b_{k-1} b'_{k-1}} , & \text{for} \quad k = 1, 2, ..., K - 2 \end{cases} \tag{3.7}$$

and

$$d'_k = \begin{cases} \frac{d_0}{a_0} , & \text{for} \quad k = 0 \\ \frac{d_k - b_{k-1} d'_{k-1}}{a_k - b_{k-1} b'_{k-1}} , & \text{for} \quad k = 1, 2, ..., K - 1 . \end{cases} \tag{3.8}$$

Back-substitution yields the desired values

$$x_k = \begin{cases} d'_{K-1} , & \text{for} \quad k = K - 1 \\ d'_k - b'_k x_{k+1} , & \text{for} \quad k = K - 2, K - 3, ..., 0 . \end{cases} \tag{3.9}$$

The algorithm is not generally stable, but stable for diagonally dominant matrices [Dat10].

Introducing image diffraction edges (modified ESEA)

Theoretically, the solution to the ESEA delivers the desired result. However, a more efficient approach is proposed by introducing diffraction into the image model using *image edges*. For each specular reflection, subsequent diffractions are considered in the *mirrored space*, instead. In this way, every reflection is removed and the otherwise heterogeneous path finding routine is reduced to a homogeneous diffraction problem operating on a specific set of regular edges and image edges. The cost of a mirror transformation of edges is traded for a simplified scheme to determine the values s_k of the parametric apex point representation, see Equation (3.10). The approach exploits the distance preservation property, making a reverse transformation of apex points found in the mirror space obsolete.

As also necessary for the traditional MIM, a visibility test must be performed that validates three conditions:

- A reflection point must lie inside the two-dimensional polygon boundaries,
- a diffraction apex point must lie within the one-dimensional edge boundaries and
- the free path between any two interaction points must not intersect any other polygon.

The angle terms are *non-linear* as the Euclidean distance normalisation is required to apply their scalar product in the formulation of the angle constraints. Hence, rewriting the matrix coefficients results in a system with non-linear terms, which is undesirable for fast calculations. Instead, a linearisation of the normalisation value is proposed and an iterative routine approximates the analytical interaction point instead to keep run-times manageable.[18]

Considering that reflection points are substituted by an image representation, only diffraction constraints will finally remain. To implement a fast routine, it is helpful to use a *parametric* representation of the apex (interaction) point I_k

[18] Depending on the implementation and the number of iterations used for the approximation, slightly different results may be obtained. Especially for numerically unstable constellations, the deterministic property of the method is violated, which should be kept in mind.

with respect to the vertices $P_{k,\text{start}}$ and $P_{k,\text{end}}$ of the diffraction edge. Assuming finite edge lengths, as found in a polygonal mesh representation, this yields

$$\begin{aligned} I_k(s_k) &= P_{\text{start}} + s_k \cdot (P_{\text{end}} - P_{\text{start}}) \quad \Leftrightarrow \\ &= (1 - s_k) \cdot P_{\text{start}} + s_k \cdot P_{\text{end}} \end{aligned} \tag{3.10}$$

and an image that has been mirrored by l preceding reflections

$$\begin{aligned} I_{k,l}(s_k) &= P_{l,\text{start}} + s_k \cdot (P_{l,\text{end}} - P_{l,\text{start}}) \quad \Leftrightarrow \\ &= (1 - s_k) \cdot P_{l,\text{start}} + s_k \cdot P_{l,\text{end}} \,. \end{aligned} \tag{3.11}$$

The parameter $0 \leq s_k \leq 1$ defines the relative location of the apex point on the edge *and* image edge.

The iterative routine starts with an arbitrary[19] initial value $s_k^{(0)}$ for each segment k of the path. The distance between interaction points can be assumed constant while the angular constraint remains (initial value $s_k^{(0)} := 0.5$ is implemented per default). Therefore, as regards Equation (3.3), an *approximate denominator* is used instead, which is *independent* of the apex point to be determined (i.e., is constant per iteration). The k-th term of the (i)-th iteration in the equation system using the parametric apex point representation and the constant denominator, obtained from the preceding iteration $(i-1)$, yields

$$\frac{\overrightarrow{I_k^{(i)}(s_k) I_{k-1}^{(i)}(s_{k-1})}}{\|\overrightarrow{I_k^{(i-1)} I_{k-1}^{(i-1)}}\|_2} \cdot \overrightarrow{m_k} = \frac{\overrightarrow{I_{k+1}^{(i)}(s_{k+1}) I_k^{(i)}(s_k)}}{\|\overrightarrow{I_{k+1}^{(i-1)} I_k^{(i-1)}}\|_2} \cdot (\pm\overrightarrow{m_k}) \,. \tag{3.12}$$

Put simply, the approach accepts a slightly skewed wavefront vector representing the incoming and outgoing directions of the local diffraction problem. In this way, the analytical solution is estimated by solving linear equations only, and the locally correct apex points are determined instead.[20] Larger distances between segments lead to smaller errors in the direction vectors, hence the convergence speed towards the analytical solution is faster (i.e., higher accuracy, or less iterations are required). Per iteration, the scaling values s_k are corrected and the compliance

[19] If a random value is chosen, the result is non-deterministic.

[20] Again, strictly spoken, the deterministic nature of the approach is violated here as the result may be affected by the selection of the initial value of s

with Snell's law (the correctness of the local angle constraint) is re-evaluated. This procedure is repeated until either a sufficient resolution is obtained (regarding angle or distance thresholds), or a maximum iteration number is reached.

Integrating the approximation into the tri-diagonal matrix is achieved by simplifying and rearranging Equation (3.12), assuming that all reflections have been solved by mirror transformation and determination of K image edges. Consequently, only positive signs are present in the ESEA and the normalised edge direction vectors $\overrightarrow{m_k}$ are substituted by edge parameters leading to

$$\overrightarrow{m_k} = \frac{\overrightarrow{u_k}}{||\overrightarrow{u_k}||_2} = \frac{\overrightarrow{P_{k,\text{end}}} - \overrightarrow{P_{k,\text{start}}}}{||\overrightarrow{P_{k,\text{end}}} - \overrightarrow{P_{k,\text{start}}}||_2} . \tag{3.13}$$

Also, the supporting constant d_k is introduced as the inverse of the denominator in Equation (3.12), yielding

$$d_k = \left(||\overrightarrow{I_{k+1}^{(i-1)} I_k^{(i-1)}}||_2 \right)^{-1} . \tag{3.14}$$

If the vector notation from interaction points $I_{k-1}(s_{k-1})$ to $I_k(s_k)$ is replaced by the notation from Equation (3.11), the edge direction vector $\overrightarrow{u_i}$ from Equation (3.13) and the simplification in Equation (3.14) is substituted in Equation (3.12),

$$\begin{gathered} d_k \cdot \left(\overrightarrow{P_{k,\text{start}}} + s_k \cdot \overrightarrow{u_k} - \overrightarrow{P_{k-1,\text{end}}} - s_{k-1} \cdot \overrightarrow{u_{k-1}} \right) \cdot \frac{\overrightarrow{u_k}}{||\overrightarrow{u_k}||_2} = \\ \frac{\overrightarrow{u_k}}{||\overrightarrow{u_k}||_2} \cdot \left(\overrightarrow{P_{k+1,\text{start}}} + s_{k+1} \cdot \overrightarrow{u_{k+1}} - \overrightarrow{P_{k,\text{end}}} - s_k \cdot \overrightarrow{u_k} \right) \cdot d_{k+1} \end{gathered} \tag{3.15}$$

is obtained. Considering a representation that complies with a tri-diagonal matrix format, one can define

$$\begin{aligned} a_k s_k + b_k s_{k+1} &= c_k \,, \quad &&\text{for} \quad k = 0 \\ b_{k-1} s_{k-1} + a_k s_k + b_k s_{k+1} &= c_k \,, \quad &&\text{for} \quad k = 1, 2, ..., K-2 \\ b_{k-1} s_{k-1} + a_k s_k &= c_k \,, \quad &&\text{for} \quad k = K-1 \end{aligned} \tag{3.16}$$

with the scalar terms

$$a_k = \|\overrightarrow{u_k}\|_2^2 \cdot (d_k + d_{k+1}) \,, \tag{3.17}$$

$$b_k = -(\overrightarrow{u_k} \cdot \overrightarrow{u_{k+1}}) \cdot d_{k+1} \tag{3.18}$$

and

$$c_k = -(\overrightarrow{P_{k-1,\text{start}}} - 2 \cdot \overrightarrow{P_{k,\text{start}}} + \overrightarrow{P_{k+1,\text{start}}}) \cdot \overrightarrow{u_k} \,. \tag{3.19}$$

In a matrix notation, the target vector s represents the scaling values to obtain the apex point between the start and end point of each diffraction edge in the original Euclidean space. Solving

$$A^{(i)} \cdot s^{(i)} = c^{(i)} \tag{3.20}$$

with respect to the matrix A, consisting of the scalar terms calculated by Equation (3.17), Equation (3.18) and the vector c from Equation (3.19) determines the apex points of iteration (i) based on values of the previous iteration or the initial values of s. The apex points $I_k^{(i)}$ are evaluated with the values of s_k according to Equation (3.13). Note that the variables s_k are only used for the next iteration in the approximated denominator term, cf. Equation (3.14), if the resulting propagation path length has decreased or the geometrical angles of the path segments with respect to the corresponding edge do not approximate the angle constraint sufficiently. Since a set of apex points represents the shortest path via a specific set of edges under the assumption that each segment complies with the far-field condition, the solution converges towards the global minimum of the overall path length. A range validation of s_k performs the visibility test for the diffraction boundaries without the need to resolve the apex location yet (i.e., $0 \leq s_k \leq 1$ is assessed). However, the reflection points must still be determined and validated. This is done afterwards in a segment-by-segment fashion for every propagation path section, wherein a mirror transformation must be applied. If one intersection test fails, the entire path is neglected, otherwise the path is marked as valid.

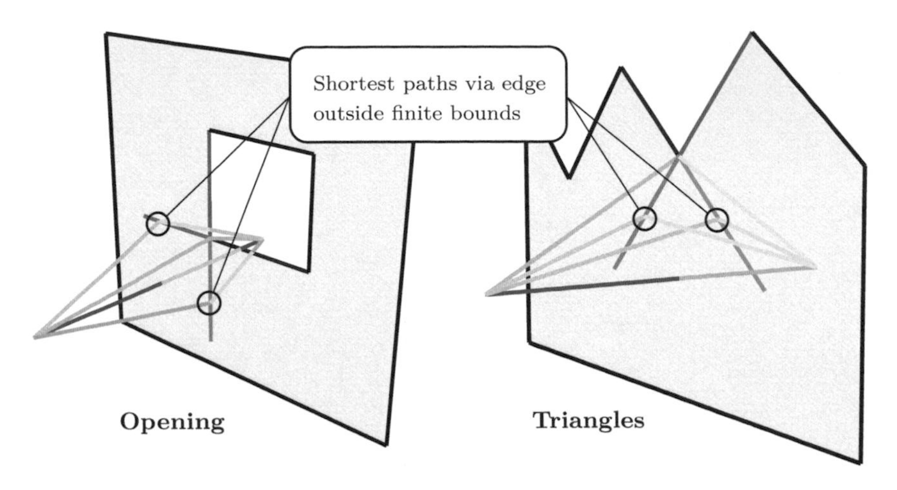

Figure 3.15: Scenarios where the method fails to find a propagation path, because the shortest paths via edges are out of bounds.

3.6.3 Compatibility with acoustic propagation models

The IEM solely generates candidates following the shortest path along a wavefront normal via a set of edges or images of edges without considering the edge's finite size and the adjacent edges of start and end vertices. The visibility test then verifies whether the path candidate crosses elements within valid ranges and removes paths that do not satisfy any of the constraints. Strictly speaking, this intervention prematurely assumes, that no diffraction occurs if the shortest path via the sequence of elements is not available. However, it is conceivable that a diffraction model (such as the BTMS, cf. Section 2.3) is able to handle finite diffraction contributions via an edge with an apex point lying outside the edge boundaries. It is the objective of the propagation model to determine the acoustic content of each geometrical path. Nonetheless, the validity of the employed acoustic models correlate with the geometrical pathfinder and, in the final analysis, both must be compatible to withstand a benchmark test. The modelling topic is described in detail in Section 3.5 and a validation is presented in the benchmarks, see Chapter 5.

At this point, the compatibility of the pathfinder with the acoustic modelling of the two phenomena reflection and diffraction is inquired. As a GA approach does not consider waves in the physical sense, but treats sound as rays or particles, the circumstances in the vicinity of a geometrical path are unknown and must be assumed isotropic for certain spatial dimensions. For example, the reflection model expects that the wave type is not altered – a spherical wave remains a spherical wave – but an angle-dependent and frequency-dependent attenuation can be readily integrated. Except for the Maekawa model [KP74], diffraction models account for the modification of the wave type *after* interaction with the edge (a spherical wave becomes a combination of spherical and cylindrical wave type), but assume an otherwise equal behaviour of adjacent wavefront normals.

If geometrical paths intersect in close proximity to boundaries of finite elements the prerequisite of similar behaviour in the vicinity is clearly violated. The magnitude of error commonly scales with the ratio of relative distance and wavelength, but some transitions affect the entire frequency range, for example the transition at reflection and shadow boundaries (cf. Figure 2.5). A critical review of GA methods reveals that disregarding wave behaviour generally promotes discontinuities caused by the finite sizes of geometrical elements of virtual environments beyond canonical situations. Hence, a concept is required to deal with such discontinuities. It is believed, that the objective of the IEM to find all possible shortest paths via the geometrical shapes including relevant edges partly tackles this problem. Based on the reasoning presented in Section 3.5, it can be concluded that the sum of valid geometrical paths with any reflection and diffraction combination delivers sufficient data to model a continuous sound field to a satisfactory degree for urban auralisation purposes. In anticipation of the diffraction model's feature requirements, the UTD approach is able to literally *fill the gaps* of discontinuities evoked by reflections at finite-sized polygons considering the path contributions along the polygon's edges. The continuity principle requires to cover both diffraction into the shadow region and into the illuminated region, because each component is responsible for the compensation of the energy mismatch in the absence of diffraction (cf. Section 2.3.2).

Nonetheless, situations can be envisioned in which the proposed pathfinder fails to deliver the shortest propagation path, for example, through an inconveniently shaped opening or a raster of steep triangles. A selection of scenarios that depict difficulties of the method are shown in Figure 3.15.

In summary, the characteristic of the IEM delivers valuable results for diffraction models that require parameters of the shortest path via the respective edge, such as UTD and Maekawa. Geometrical objects that mainly consist of convex shapes are accurately determined, for example, buildings with traditional roof shapes. Situations can be found, where the presented implementation does not return a propagation path, for example, if a path slides into a corner of two wedges in an

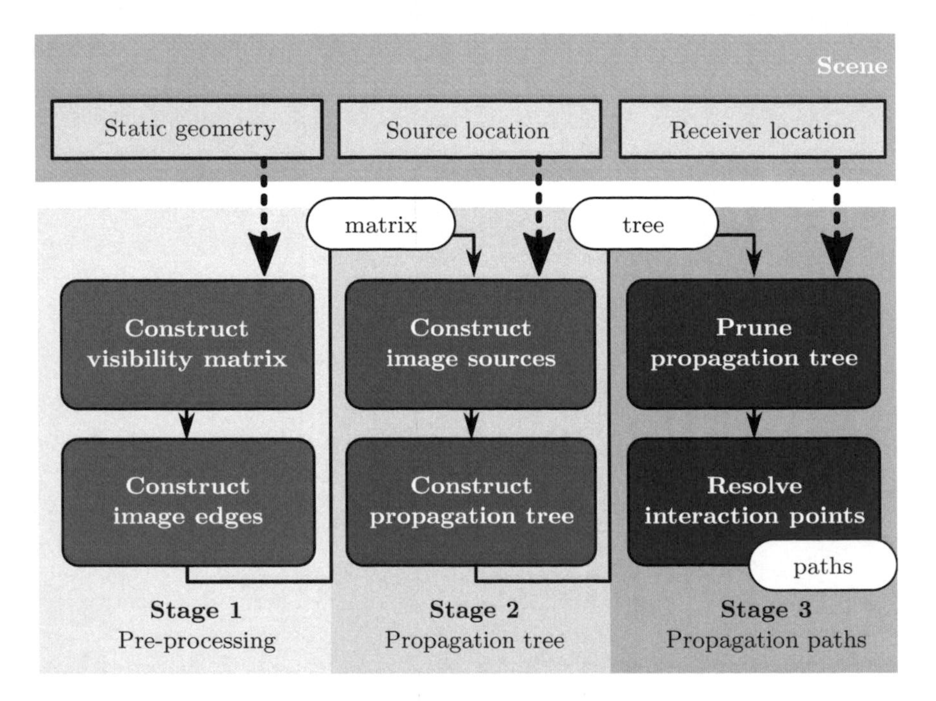

Figure 3.16: Pathfinder routine block diagram.

opening or steep triangle valley. Anyway, these constellations do not comply with the underlying idea of a shortest path in the first place, and a solution requires sophisticated integral diffraction models, like BTMS, a topic that is addressed in the conclusion, see Chapter 6.

3.6.4 Approach

To implement the pathfinder algorithm featuring the determination of geometrical propagation paths with an *arbitrary sequence* of reflections and diffractions, a division of the approach into three main stages is suggested:

1. Inter-mesh visibility matrix and image edge construction.
2. Propagation tree construction.
3. Propagation path generation.

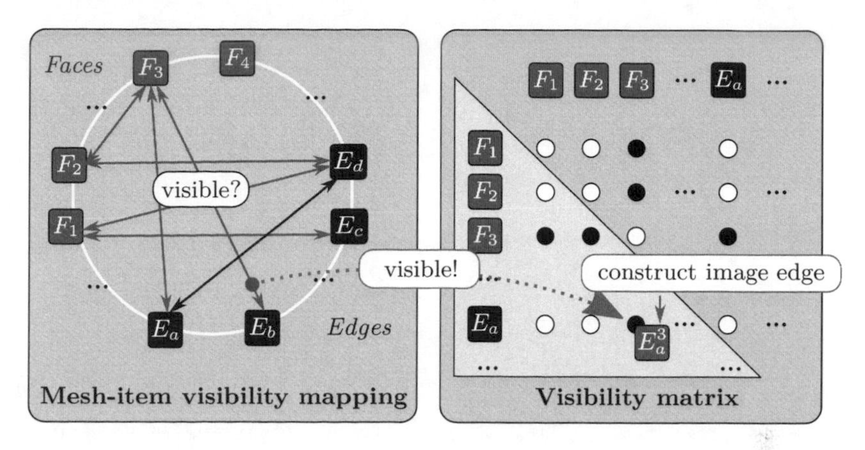

Figure 3.17: Visibility mapping and matrix construction with image edges.

The block diagram in Figure 3.16 shows the execution order, indicates the required input data for each subroutine and also names intermediate results. In the context of real-time auralisation, only the image source construction and the propagation path generation is performed repeatedly, while the inter-mesh visibility matrix and the constructed image edges are pre-calculated (only valid for static geometries). Each subroutine is shortly described to help understand its general purpose. An in-depth explanation including implementation details is given in the following subsections.

Stage 1) The pre-processing of geometry allows to generate a-priori visibility indicators among all mesh items (polygons and edges). This inter-mesh visibility matrix is established in preparation of an ordered tree data structure with the sound source as root node, referred to as *propagation tree* in the following. It is intended to speed up the construction of propagation paths. Additionally, images of the edges can already be determined at this stage. Because the location of edges is fixed with respect to the reflection plane, the vertex points are transformed and the image edges are stored for a fast look-up in the subsequent routines (cf. Section 3.6.5).

Stage 2) When the location of the sound source is known, a selection of mesh items visible to the origin is determined in the fashion of a *from-entity scanning* [Ant+12]. Furthermore, image sources can be generated. The procedure constructs the propagation tree with the root node being the sound source and the first order of branches being the visible mesh items, as depicted in Figure 3.18.

The tree is iteratively expanded and subsequent branches are inserted, if visible according to the visibility matrix, until the user-defined maximum order is reached (cf. Section 3.6.6).

Stage 3) The moment the receiver location has been given, geometrical propagation paths can be determined. To do so, the propagation tree is pruned by erasing all leafs that are not visible to the receiver, again using a from-entity scanning. The residual tree structure holds propagation path *candidates*. With both the source and receiver location at hand, the reflection points and diffraction apex points can be determined by solving an equation system including the angle constraints that lead to the shortest paths via the diffraction edges. A complete matrix inversion is avoided by exploiting the Euclidean transform properties (preservation of distance and relative angle) as well as the symmetry and tri-diagonality of the matrix (cf. Section 3.6.1). Finally, the locations of the diffraction apex points and reflection points are calculated and the path is constructed segment by segment (cf. Section 3.6.7).

3.6.5 Geometry pre-processing for static urban environments

Generating a-priori knowledge is a central concept to shift time-consuming calculations from critical procedures into a loading stage. By doing so, the possibility to modify input data required by the pre-calculations is sacrificed. In the context of urban sound auralisation, it is suggested to declare the built environment as non-modifiable during runtime and use static geometries for buildings, roadways, noise barriers, city furniture, etc. This limitation enables propagation simulations between sound source and receiver locations that can be performed fast enough to approach real-time update rates, as discussed in Section 4.4.2.

For urban sound propagation simulation, a pre-processing step is proposed that analyses the geometry mesh and generates a pair-wise inter-item relationship matrix. Such an approach is likely to impose extensive computational load and is particularly memory-intensive. In a worst-case scenario, a cost function of $F = (4N) \cdot (4N - 4) = 16N(N - 1)$ is expected for a triangulated mesh of N polygons that constructs three diffraction wedges with adjacent triangles. Without the employment of spatial data structures such as binary space partitioning, a complexity of $O(N^2)$ is found – and the equal scale applies for the allocated memory. For example, a mesh of $N = 100,000$ triangles using a connection object with a 100 Byte memory footprint requires a total amount of 4 GByte allocated data for the two-dimensional matrix construction. Therefore, in the current state, only moderately sized environments are feasible and can be used for real-time rates.

In the context of acoustic propagation simulation, it is productive to provide an inter-item visibility map. It is suggested to analyse the geometry considering the *likelihood* of items constructing a valid reflection or diffraction path segment between each other. The delicate part in this approach is the lack of knowledge on the source and receiver location, which will only be available for the actual simulation routine. For this reason, all consideration must produce robust results with respect to arbitrary source and receiver locations. A first implementation of this concept aims at ruling out connections that certainly never lead to a valid path segment, regardless of the source and receiver location. For instance, a reflection at a polygon can only construct valid propagation segments with items that are located in front of the plane. Furthermore, diffraction at edges only constructs valid propagation segments with items that are located outside the wedge's solid volume and therefore items that entirely lie behind both wedge planes are invalid. This kind of back-face culling is able to approximately cut down the visibility map entries by half for reflecting polygon items, and roughly another quarter for diffracting edge items assuming box-shaped buildings with right-angled corners. Another idea requires the relatively computation-intensive from-region visibility tests, a routine that is extensively used in beam-tracing algorithms [Ant+10; Ant+12]. Considering a connection between one mesh item and another, every possible direct connection may be occluded by other objects and ergo are not able to construct a valid propagation path. The implementation, however, requires sophisticated scanning methods, imposes conditions on the shape of polygons (i.e., convex shapes, usually triangles) and may only be feasible with underlying spatial data structures offering fast iteration routines.

For urban environments, the constructed matrix is expected to be sparse, bearing in mind that buildings or street canyons have a limited visibility region. Entries that are not able to establish a valid subsequent propagation path can be omitted, saving memory allocation. For example, buildings represented by convex cube-shaped bodies are not able to establish valid self-referring reflections, and each triangle can only construct (higher-order) diffraction propagation paths with adjacent triangles over the three edges. However, non-ideal scenarios exist where these assumptions do not hold, for example, the internal area of a large detailed square.

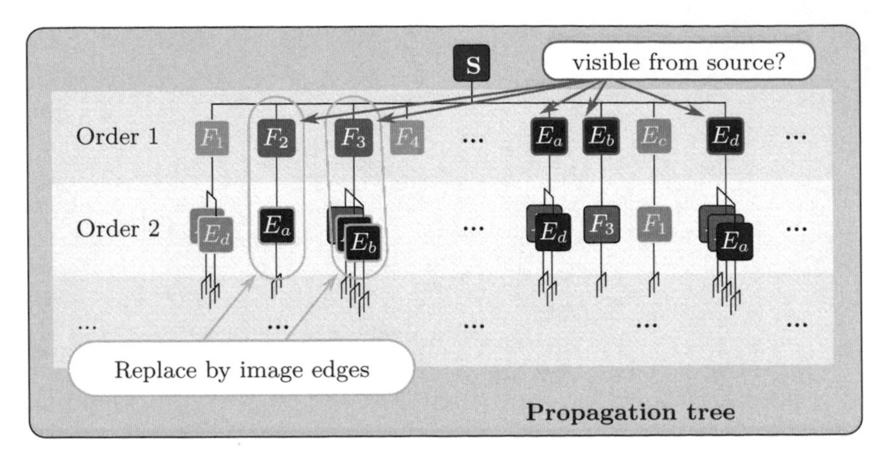

Figure 3.18: Propagation tree connecting visible mesh items from the source location (root node). Edges that follow after reflections are replaced by image edges.

More considerations to prune the visibility matrix are imaginable, for example, by evaluating against thresholds that include estimations of sound propagation attenuation either based on actual models or statistically established penalty values, which naturally require a validation for the desired city model. In anticipation to employ such methods in the future, a configurable, flat penalty attribution for spherical spreading loss, reflections and diffractions has been realised in the IEM implementation in order to provide this feature for later use. Links between far-distant mesh items are evaluated against a *dynamic range value* (which can be set to infinity). If exceeded, the corresponding mesh items are not able to establish a propagation coupling.

After the determination of the inter-item visibility matrix, a fast *traversal* of mesh items that potentially form valid geometrical propagation paths is at hand.

3.6.6 From-entity region scanning

The pathfinder algorithm begins with the determination of visible mesh items in the vicinity of the sound source location. From-entity region scanning is performed by iterating over all mesh items and evaluating the possibility of a sound propagation connection applying similar methods as in the aforementioned geometry pre-preprocessing (cf. Section 3.6.5). First, a face-normal check is performed

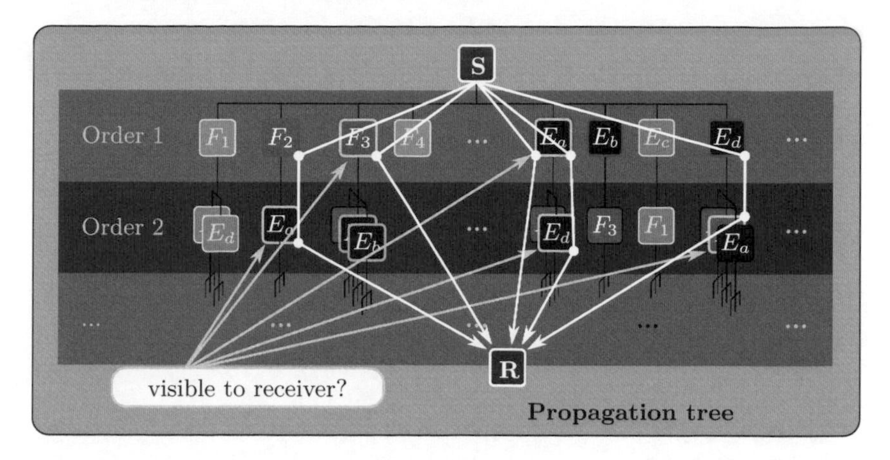

Figure 3.19: Propagation path candidates are established by connecting nodes visible to the receiver.

to ascertain that reflections at polygon items and diffractions at wedges are within valid volumes. The set of residual mesh items are regarded as branches of a tree structure with the sound source being the root node. The preparation of the tree structure is important to identify mesh items that represent seeds for further propagation paths. Hence, edges that follow after faces in the tree can already be replaced by the corresponding image edges.

In the same fashion, a set of mesh items is determined for the receiver location.

3.6.7 Path linking

After a subset of mesh items has been found that can be directly connected to either the sound source or the sound receiver location, a path linking routine must be performed. This part can be seen as the central element of the *image-edge pathfinder algorithm.*

Being part of the third stage in Figure 3.16, the construction of geometrical propagation paths requires the determination and validation of acoustic interaction. The approach begins with a traversal of the propagation tree with the intention to find sequences that are visible to both the source and the receiver. At this stage, propagation paths are represented by a list of mesh items that are *potentially* creating a valid path. Therefore, they are recognised as *candidates.* Without applying any acceleration methods, each candidate must be geometrically validated by a visibility test. However, because the sequence contains surfaces and edges, either the classical ESEA must be solved, or all diffracting edges succeeding a reflecting surface are substituted by an *image edge* in order to solve the more efficient *modified Equation System of Equal Angles* (mESEA) that only contains diffraction terms (cf. Section 3.6.2). During runtime, the routine evaluates the validity of the apex point of every diffraction. Only if all interaction points are inside the bounds of the finite edges is a subsequent visibility test performed. If a path candidate is found valid, a second routine verifies whether the reflection points intersect the corresponding polygon and are not occluded by other geometries. This step requires the transformation of the images generated by the mirroring method into the original space. By an iteration over the segments of the propagation path under examination, apex points are transformed and reflection points are calculated by a line-plane intersection algorithm, until the final path in the initial space (world coordinates) is restored. After this final step, the geometrical propagation path has been found and can be passed on to the acoustic modelling, as introduced in Section 3.4.

Pruning the tree structure *during construction* and intervening *during validation* is highly effective and user options have been integrated that allow to activate these methods bearing real-time auralisation in mind. However, some risks are implied by using abortion criteria and progressive acceleration methods that are discussed in Section 3.6.8.

3.6.8 Abortion criteria and progressive acceleration methods

Abortion criteria configure the runtime and quality of the propagation simulation. Acceleration methods are options to optimise the execution. The specific parameter selection of both measures is context-dependent and commonly chosen to match a certain purpose based on personal experience, perceptual verification (e.g., informal listening tests) or using other evaluation procedures. Naturally, for real-time auralisation, these parameters must be pushed to the limits in order to achieve sufficient update rates – even if powerful computer capacity is at hand.

The classical abortion criterion that is basically always in place for mirror image methods is *order capping*. The maximum number of reflections and/or diffractions, the most influential parameter on runtime and memory consumption, is considered *fixed*. Due to the exponential complexity and data grow rate, already orders greater than 2 can result in extremely challenging requirements – even for moderately complex environments. Allowing only reflections and/or diffractions of first and second order is an effective way to maintain a relatively good performance, usually making real-time update rates possible for non-optimised algorithms. However, from an acoustic perspective, such low orders have the detrimental effect to ignore important propagation paths that require higher orders, predominantly occurring between parallel walls or street canyons. Hence, order capping can only produce an adequate resolution for environments that do not require multiple reflections and diffractions in order to find the perceptually most relevant propagation paths.

In contrast, sophisticated algorithms that can handle higher orders and apply progressive acceleration methods to maintain an acceptable runtime and memory footprint have the potential to implement a routine that is closer to a perception-driven approach. However, such progressive methods must apply technical or perceptual assumptions, which in turn must be verified. In the course of this thesis, several ideas have been realised for future use that accelerate the pathfinder algorithm significantly, but a general applicability cannot be provided and must be proven individually.

The first method is called dynamic range culling and is based on the physics-based sound effect of spreading loss (cf. Section 3.4). Either, a distance attenuation level is subtracted from the initial sound power level of the source during evaluation, or an absolute threshold level is determined to dismiss paths with an exceeding attenuation level. It is considered that sounds below a certain level threshold are impossible to be detected, which is evaluated by the geometrical distance of entire propagation paths or path segments. Therefore, a pruning routine conservatively estimates an item-to-item distance value during the construction of visibility and discards perceptually negligible connections to increase the visibility matrix sparseness. In addition, the distance evaluation is also applied during path construction, which is effectively accelerating the sorting process by making further and more computationally expensive evaluations obsolete.

In fact, progressive approaches that estimate the perceptual relevance based on propagation models, without actually calculating the costly acoustic properties are highly effective. For that reason, further very simple but extremely powerful options have been realised, for example, a single-value (broadband) level penalty accumulation routine that is applied in the same manner as the dynamic range culling with the difference that a pre-defined level per reflection and diffraction is subtracted from the initial sound power. The definition of the penalty values can be based, for example, on a statistical analysis of data obtained through high-resolution simulations of the given environment.

Another accumulating method accounts for the diffraction angles and dismisses propagation paths that exceed a user-defined maximum diffraction angle, as it is the case for a propagation paths taking a complete circular round-trip at a building. Furthermore, back-scattering can be completely deactivated, which is a helpful acceleration method if most wedges fail to establish a shadow region. However, this method is a questionable option for dynamic urban environments, since the back-scattered diffraction energy plays an important role for the continuous sound field simulation (cf. Section 2.3.2).

To speed up simulation rates for real-time purposes, the solution of the mESEA using the approximating iterative approach can be parametrised. An error threshold for the angle constraint can be specified to ensure the desired simulation accuracy. Also, a maximum number of iterations can be chosen to limit the calculation runtime. Although not quantifiable, the expected inaccuracies are not likely to be detrimental.

Finally, the very costly free-path visibility check can be disabled individually for the from-region scanning of the source, the receiver and for intermediate segments of a propagation path. Although it is an important feature, it is also excessively costly – unless spatial data structures are employed that significantly accelerate the intersection test routine (which have not yet been realised). Particularly for simplified scenarios (street canyons, squares, etc.), the segment-wise validation between reflections and diffractions are often unnecessary and represent an option to increase the simulation rate for a real-time auralisation application.

3.7 Rotation-based pathfinder algorithm for convex meshes

The challenge to find high-order diffraction paths gains importance in scenarios, where an obstacle blocks the direct sound path and a single-diffraction handling is not sufficient to determine any geometrical path around the occluder. This can easily happen if a building, bus stop cabinet, a tree trunk or a truck with box-like shape is in close vicinity to the sound source or the sound receiver, as depicted in Figure 3.20.

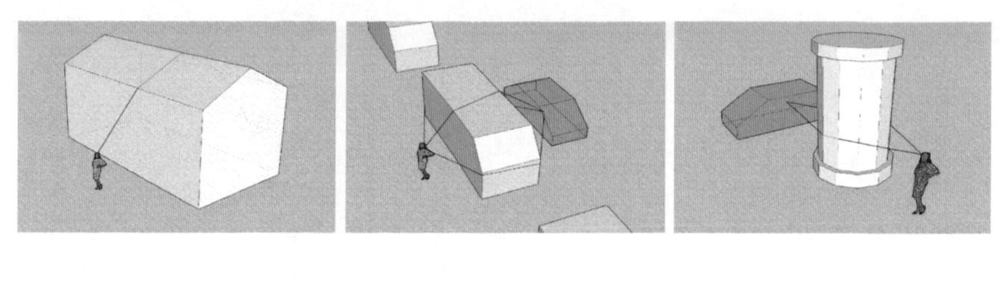

(a) Gable roof building. (b) Parking vans. (c) Poster pillar.

Figure 3.20: Scenarios with obstacles requiring diffraction orders > 1.

Another deterministic propagation pathfinder algorithm has been proposed by Stienen and Vorländer in the context of real-time auralisation [SV16; SV17]. The initial motivation was to find propagation paths potentially containing many diffractions that bend into the acoustic shadow region of an occluder. This investigation aimed t preventing unnatural situations during auralisation outdoors, where there is absolute silence from an otherwise dominant sound source, for example, from a vehicle passing a building or a row of parking cars or trucks. Additionally, it appears invaluable to find the shortest route around an occluding object and combine the result with a simplified acoustic model for diffraction, such as the Maekawa method. Therefore, reflections have been neglected for the benefit of a reduced problem that exclusively consists of subsequent diffractions. One can imagine that the algorithm behaves like a crawler that spins a yarn from the source around the occluder to the receiver (or vice versa) – or a set of rubber bands that are spanned across the object slipping into the path of least stress. A prerequisite of the approach is that it must only incorporate lightweight calculations (e.g., no matrix inversion) in order to deliver simulation results for high-order diffraction problems at interactive rates to maintain applicability for a real-time auralisation.

For mathematical correctness on the one hand, and simplification reasons regarding the implementation on the other hand, the presented algorithm is only considering convex shapes. However, it is conceivable to extend the algorithm to handle non-convex forms. It is further required that the underlying data structure inherits links between polygons and edges, such as a coherent polygonal mesh using half-edges [Bot+02], as a feature to find edges connected to an illuminated face which is required, as described in the following.

The approach is rooted in the idea that the angle constraint of the shortest path via a set of diffraction edges is maintained if the source or receiver entity is *rotated*

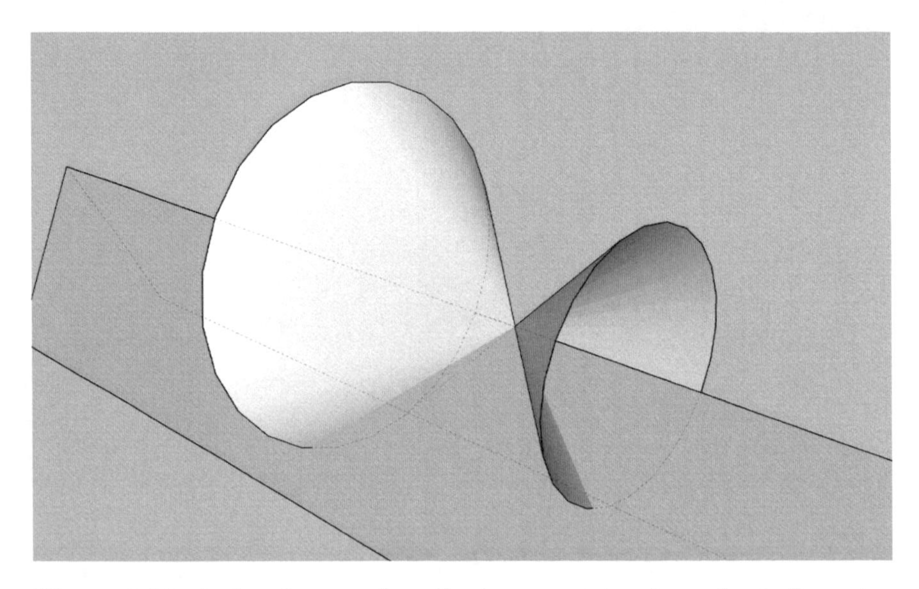

Figure 3.21: Ambiguity cones for a fixed apex point and wavefronts (incoming and outgoing angles) for a wedge-type diffraction.

around the edge axis. In fact, an apex point on the edge and a pair of incident and outgoing directions inherits a certain ambiguity regarding the source and receiver locations [Tsi+01], as shown in Figure 3.21. Applying to both entities, the location can be freely moved on the surface of a cone with circular base and its rotation axis along the edge. The cone's parameters are determined by the diffraction problem. The opening angle corresponds to the angle that is formed by the incoming and outgoing wavefront direction vector and the edge direction. The tipping point coincides with the apex point on the edge.

This feature is exploited to reduce the complexity of the diffraction problem by one dimension. This can be achieved by transforming the locations using a rotation operation and project them into a virtual plane that also contains the edge. This routine is performed on a sequence of edges beginning at the source and finalised at the receiver (however, it can be realised vice versa without limitation). The segment-wise rotation of the source is continued until reaching the plane that is spanned by the last edge and the receiver location. In a reverse fashion, apex points are determined by parameters established in the reduced space (the two-dimensional plane) segment by segment, until the actual source location is reached.

(a) Scenario with receiver (blue) and source (red).

(b) Rotation #1 into plane of right-hand roof polygon (red circle).

(c) Rotation #2 into plane of left-hand roof polygon (green circle).

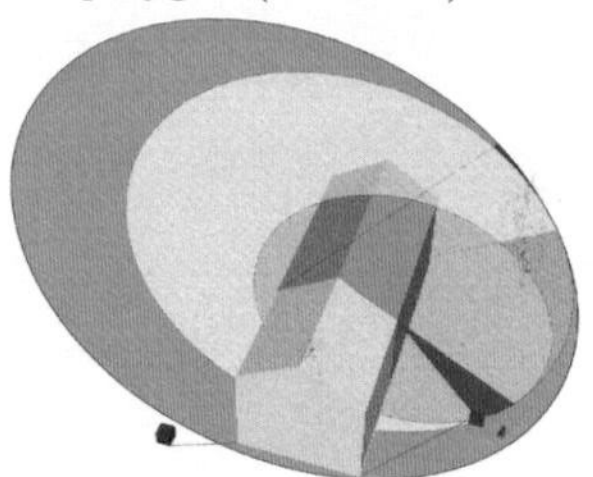

(d) Final rotation into plane of last edge and receiver location (blue circle).

Figure 3.22: Rotation-based diffraction pathfinder animation: dimension reduction.

An animation is depicted in Figure 3.22 for a diffraction problem via three edges of a saddle roof building. In the first step, the source is rotated around the edge, constructed with the right-hand roof polygon and the right building facade, into the plane of the roof polygon, as indicated in Figure 3.22a. The edge circle of the light-red disc represents the rotation freedom of the source that maintains the angle of incident of the wavefront with respect to the edge, as requested by the approach. The dark-red circle element is the actual rotation distance required to poject the source location into the corresponding edge face, the right-hand roof plane in this case. Considering this source image, a straight line can be drawn from the rotated location across the next edge formed by the roof's gable. This second edge is again used to define a rotation circle, as indicated in Figure 3.22b as a light-green disc with the rotation element in intense green. The initially transformed source image is further rotated around the edge axis into the plane of the left-hand roof polygon that also includes the last edge formed by the roof and the left facade. The last step requires to perform a final rotation to move the image of the source location into the plane of the corresponding edge (roof

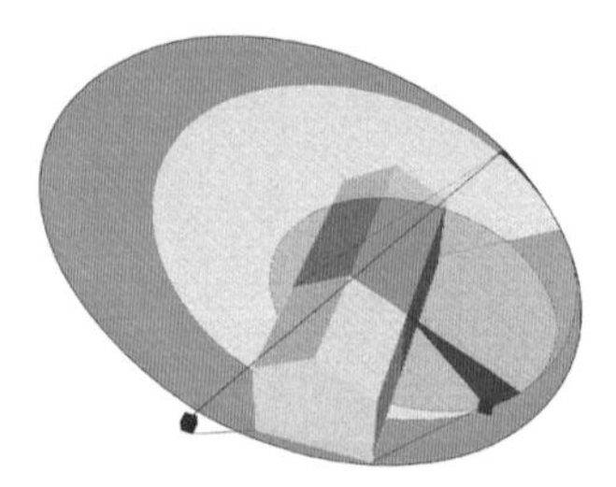

(a) Receiver (blue) and final rotation source (top-right on blue disc) are in line-of-sight touching the last edge's analytical apex point (blue line).

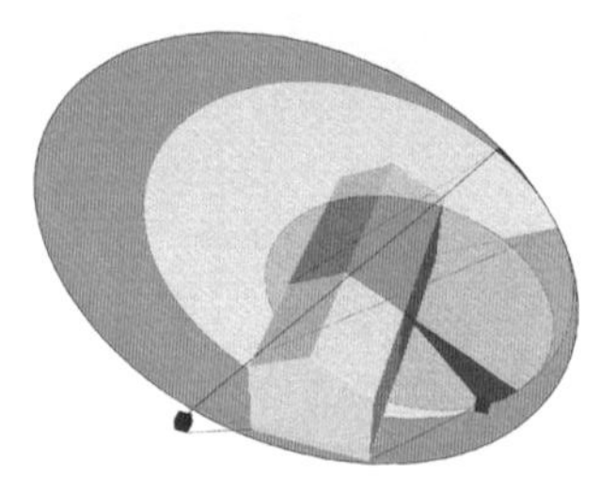

(b) Reverse substitution #1 using previous apex point and rotated image (green line) to determine the next apex point.

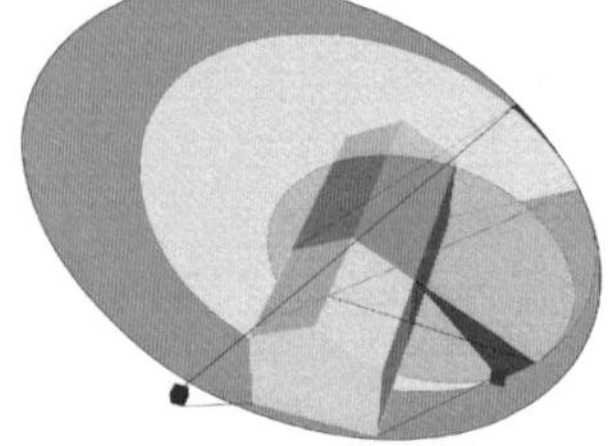

(c) Reverse substitution #2 using previous apex point and rotated image (red line) to determine next apex point.

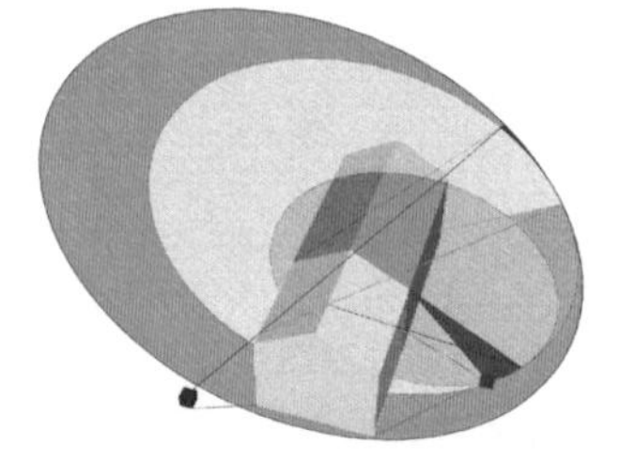

(d) Final apex point determination using previous apex point and source location (orange line).

Figure 3.23: Rotation-based diffraction pathfinder animation: reverse substitution.

and left facade) and the receiver location. This step is shown in Figure 3.22d as a blue disc. The receiver is now in *direct* line-of-sight with the transformed source, touching the analytical apex point of the last diffraction, as indicated by the solid blue line in Figure 3.23a.

The apex point is determined by solving a line intersection equation by linear algebra in a two-dimensional space and applying the resulting parameters to the three-dimensional vector representation. This procedure is repeatedly executed for each newly determined apex point with the previous, rotated source location in order to subsequently resolve all the apex points on the edges. As soon as the last edge is evaluated, a canonical situation is at hand with the peculiarity that one entity (the previous apex point) is located on a reflection plane of the wedge. The last apex point seals the propagation path as now all interaction points have

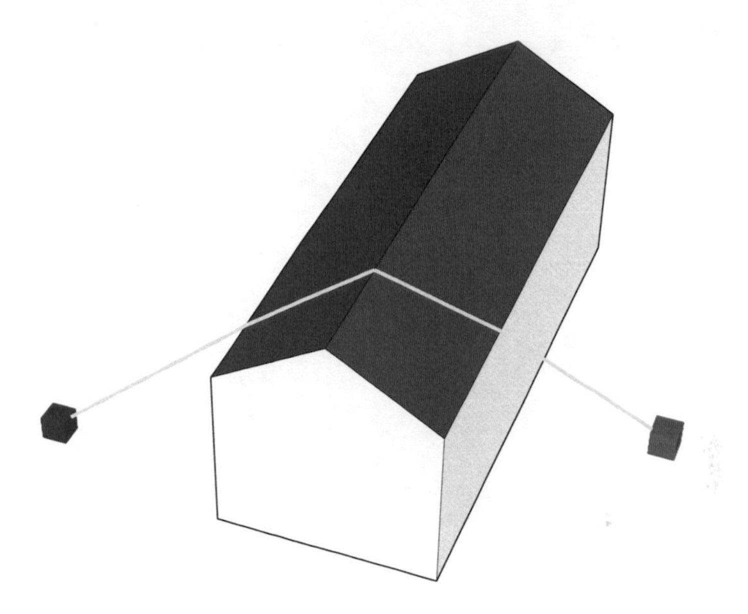

Figure 3.24: Final propagation path over a gable roof building determined by the rotation-based diffraction pathfinder algorithm.

been determined. The final path, visualised in Figure 3.24, is in line with the analytical solution of the previously mentioned general ESEA (cf. Section 3.6.2). The propagation path obtained via the sequence of edges under evaluation is deemed valid if the apex points are located on the finite edges, and otherwise dismissed.

An aspect that has been disregarded so far is related to the question, how a sequence of edges representing a path candidate is found in the first place, and what abortion criteria are conceivable. A similar concept as proposed in the previous section can be used, aiming to establish visibility metadata for mesh items such as edges and faces (cf. Section 3.6). For the rotation-based pathfinder algorithm, however, a face illumination routine is preferred that executes a from-entity scanning by marking faces that are visible to the source/receiver as *illuminated.* Since the algorithm expects convex shapes (or hulls), an object must have at least one face visible to each entity (as long as the mesh does not contain holes). If more faces are illuminable a *contour* can be detected that separates the illuminated area from the shadow area on the mesh.

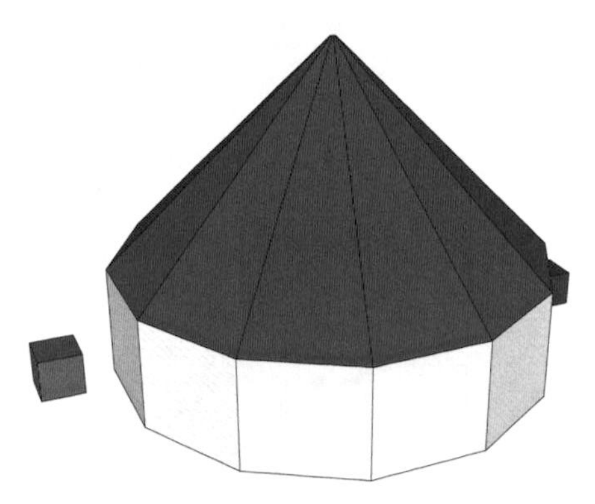

Figure 3.25: Unconventional building with a source (red) in the front and a receiver (blue) located behind the object.

A scenario that reflects this line of reasoning is depicted in Figure 3.25. An unconventionally shaped building with a yurt-like roof is considered. A source and receiver are placed in the front and back of the object, respectively, at relatively close distance. These two locations each illuminate one facade of the building, as indicated by the intensive blue polygon in Figure 3.26a and intensive red polygon that must be imagined at the backside of Figure 3.26b. Beginning at these illumination regions, a routine is traversing over the mesh by iterating over the face's edges and assigning illumination orders to every adjacent, non-marked face. The progression is visualised in Figure 3.26a and Figure 3.26b by increasingly darkened colours. Evidently, an illumination order abortion criteria is selected that stops the routine. In this case, after four edge crossings the procedure is interrupted for a clear visualisation (note, that the mesh has been triangulated, so the facade circling over otherwise rectangular shapes is deliberately inhibited). A diffraction order of $4+4=8$ is required to determine the first propagation path candidates in this scenario. The faces that are reachable by both illumination progression threads mark the mesh items that must be included by each sequence of polygons/edges leading to a propagation path candidate (see purple polygons in Figure 3.26d). Hence, two individual tree structures of adjacent edges are established, one for the source side and one for the receiver side. Both trees begin at the intersection polygons (purple faces) and branch out to the adjacent edges (connected by the half-edge's opposite face). During traversal, the routine takes into account that only those sequences of edges are valid that progress towards the illuminated region. If an edge is encountered that connects to a directly

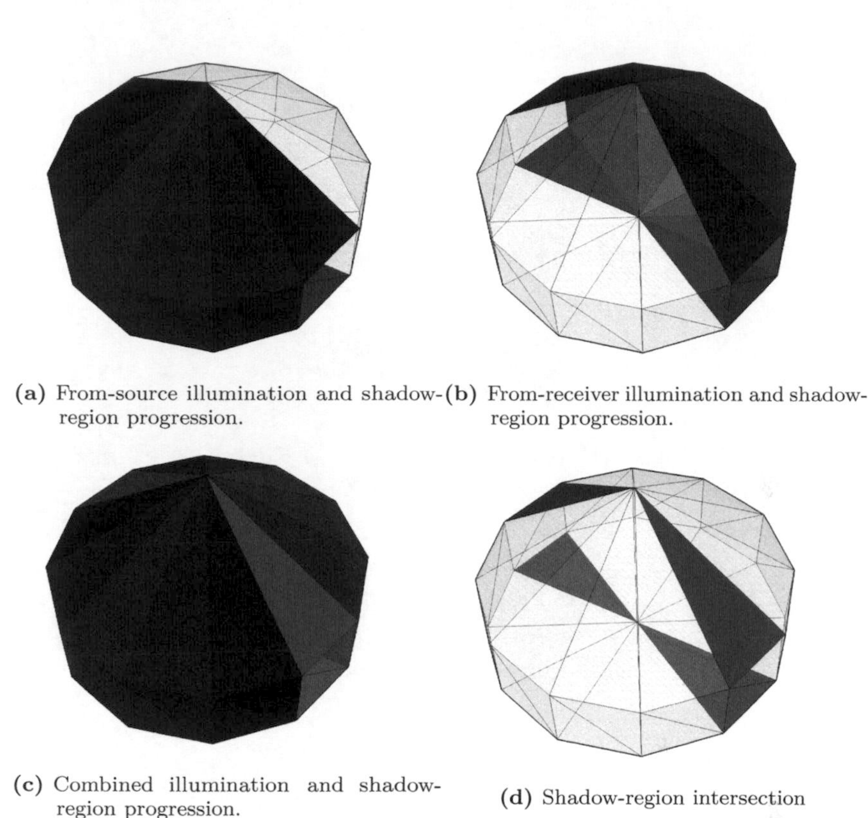

(a) From-source illumination and shadow-region progression.

(b) From-receiver illumination and shadow-region progression.

(c) Combined illumination and shadow-region progression.

(d) Shadow-region intersection

Figure 3.26: Rotation-based diffraction pathfinder animation: path candidate detection.

illuminated face, this sequence represents one half of a complete path candidate. The procedure must be performed for both branches. Finally, the combined result of both sets establishes the final list of path candidates. Although the combination possibilities seem to be extensively high, only very few sequences will lead to valid propagation paths. For example, in the considered scenario, only two paths at diffraction order 8 are found valid. The two-dimensional solution space is depicted in Figure 3.27 for a representative path.

As the rotation-based pathfinder algorithm can be rapidly executed and does not require a-priori calculations, it is feasible for real-time auralisation applications with the feature to incorporate dynamically moving occluders and dynamic ge-

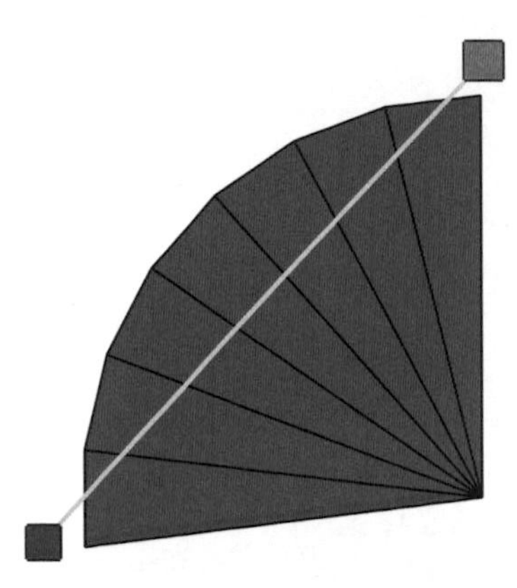

Figure 3.27: Geometrical solution in two-dimensional space for diffraction order 8 of the scenario shown in Figure 3.25.

ometries. For instance, an option to iteratively modify the height of a noise barrier and immediately perceive the auralisation result is possible at real-time rates on standard hardware. However, further concepts must be included if reflections are desired (e.g., by manually creating image sources and image receivers over a ground plane).

3.8 Summary

An auralisation concept for dynamic outdoor environments is presented that bears real-time update rates in mind and has the capability to incorporate outdoor propagation models, in particular edge diffraction and the Doppler shift. Consequently, a method based on the principle of GA is introduced as a the *pathfinder* algorithm that builds on a mirroring approach and extends the classical image models by image edges following the deterministic IEM approach. Since it is challenging to employ GA methods for dynamic scenarios regarding sound field continuity, as geometrical paths like the direct sound and reflections off walls are appearing and disappearing instantaneously over time, a solution based on the UTD model is described that effectively avoids discontinuities at critical boundaries. To achieve sufficient propagation simulation runtimes for interactive VR applications, an implementation is presented that a) requires geometry pre-preprocessing, b) can be parametrised by various abortion criteria, like a limitation of the combined reflection and diffraction order, and c) has the capacity to handle sound source and receiver updates efficiently. The algorithm is integrated in the *pigeon* application, which is part of the open-source project *ITAGeometricalAcoustics*[21]. The resulting geometrical paths can be transformed into TFs by acoustic modelling using the MATLAB® class *itaGeoPropagation*[22], as demonstrated in Chapter 5.

For constellations that require many subsequent diffractions in order to determine relevant propagation paths, another *pathfinder* algorithm that is based on a rotation principle around edges is presented, which is limited to convex-shaped objects, since it is not able to include reflections.

[21] http://git.rwth-aachen.de/ita/ITAGeometricalAcoustics

[22] Part of the of the ITA-Toolbox, http://www.ita-toolbox.org

4 Realisation of a dynamic real-time auralisation framework

4.1 Introduction

Auralisation plays a vital role in three areas of application, namely *Acoustic Assessment*, *Virtual Reality (VR)* and *Entertainment*. The applicability is not limited to these and the boundaries are overlapping. Figure 4.1 visualises the areas and indicates by the green band that *real-time processing* is required by the entire sector of *Virtual Reality* and is progressing into the others. For an acoustician, *Acoustic Assessment* is the most evident field, because auralisation represents the next logical step of presenting and evaluating acoustic content as realistic as possible. Auralisation is a supplementary tool for the qualitative evaluation of sound transmission, as the subjective perception can be assessed. It responds to the human auditory system and therefore is open and comprehensible for everyone, as already discussed in the first chapter (cf. Section 1). The *Entertainment* sector increases to adopt VR methodologies to deliver a convincing and thrilling user experience by exploiting perception and cognition aspects. However, because the generation of effects in games and movies is not strictly bound to physical correctness (yet evoke strong emotions and an overwhelming level of immersion), this thesis is not extensively concerned with the entertainment area.

An interactive component of an auralisation system requires a low-latency procedure that operates in real-time. With some limitations, a rudimentary auralisation system can be based on pre-rendered data that is presented with an interactive (real-time) toggle feature for a comparative investigation. It can partially provide the freedom to rotate the head, for example, if stationary reproduction systems with a wide sweet spot area are used, such as VBAP and HOA. However, more sophisticated auralisation applications that involve free receiver movement and employ live motion tracking require interactive rendering and reproduction stages in order to present an appropriate auralisation output at real-time rates.

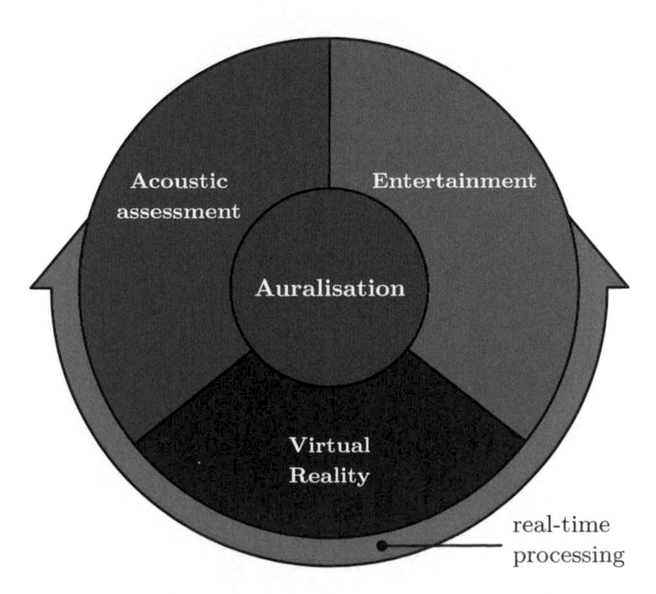

Figure 4.1: Application areas for auralisation.

The highest degree of interactivity is inevitably imposed by the application area of VR and for that reason, the real-time auralisation requirements in this thesis are dedicated to match demands raised by a VR environment. Per definition, VR requires a high level of dynamism, as it needs to respond to human interaction as realistic as possible. It gains convincing power by latency-free adaption to user input via intuitive interfaces that interact with the virtual environment (e.g., own-body user motion). In the scientific sense, a VR system has the objective to generate a virtual environment that is perceived as real and, consequently, establishes a feeling of immersion, an emotional state in which the user forgets that he/she is interacting with a technical machine operated by a computer program. Attaining this principle is challenging and can only be achieved with detailed knowledge on human perception and complex technology regarding both hardware and software. The exploitation of human perception and cognition thresholds is the prime directive for a successful VR realisation, and the simulation of physical behaviour is one of its crucial components. Transferred to acoustics, a convincing real-time auralisation system delivering audio signals to the auditory sense is indispensable, and the requirement laid upon the implementation efficiency is exceptionally high. A mismatch between expected and presented feedback in the auditory modality deteriorates the user's immersion and can lead to a complete rejection of the VR presentation.

This thesis is dedicated to the generation of real-time outdoor scenarios covering the acoustic assessment aspect by fully embracing the VR methodology (cf. Figure 4.1). The following sections describe the proposal of a real-time auralisation approach specialised for the purpose of rendering a static, outdoor environment with dynamic sound sources for an interactive user[23]. The design lays a particular focus on the implementation details for a modern Personal Computer *(PC)* architecture with a powerful Central Processing Unit *(CPU)* and high memory capacity. It aims to consistently reflect the dynamic outdoor sound propagation principles of the geometrical simulation and acoustic modelling introduced in Section 3.4 and Section 3.6 in order to integrate the methods in a real-time rendering stage. The proposed framework is following the proposed layout from Section 3.1, which is closely related to the diagram shown in Figure 3.1. The design is based on a modular object-oriented concept in which a central auralisation unit controls the entire system.

The auralisation module manages the virtual scene, delegates user input (source and receiver movements) and acts as a mediator between the simulation scheduler and the audio processor. The simulation scheduler interprets the virtual environment by means of acoustic sound propagation and delivers digital signal processing coefficients. The audio processor contains the rendering module, which runs a network of processing units reflecting the sound propagation paths as well as the directional processing engine for the receiver. The reproduction module finally prepares the rendered audio stream for the target playback system, if required. For an increased flexibility, the framework operates in different contexts and offers a control interface that can be transparently used via a network connection. The user context covers asynchronous scene updates and manages control messages. The simulation context offers a lightweight task/result communication interface[24]. It contains a sophisticated and highly configurable scheduling module, which manages a scalable number of simulation back-end instances and distributes tasks based on a user-defined decision network. The real-time audio context is driven by the audio device and is highly time-critical. It powers the digital signal network of both the rendering and the reproduction instances[25]. Additionally, motion up-sampling and parameter smoothing is performed to provide sufficiently accurate sequenced delays for a clean Doppler shift auralisation.

[23] The system is not limited to a single receiver but scales linearly with number of receivers. In a user-centred VR application, a second entity is practical to provide a fast (however unphysical) toggle mechanism for portal navigation.

[24] It was anticipated to provide a network interface between scheduler and simulation back-ends in order to enable distributed computing. However, the implementation could not be finished in time.

[25] The modular concept allows to run several rendering and reproduction modules in parallel, a feature that is valuable for prototyping, scenario toggling, etc.

Inter-process information exchange is performed with high caution to prevent any delays or blocking calls in the audio context.

4.2 Outdoor scene description

For the purpose of real-time auralisation, a virtual scene is defined by a set of dynamic objects with certain properties, a static propagation medium that transmits information by means of acoustic waves and a digital representation of the ground topology, pavements and the built environment using a static polygonal mesh.

Real-time auralisation requires fast algorithms, and GA algorithms are most suitable to apply the outdoor propagation models introduced in Section 3.4. GA methods distinguish between emitter entities (sound sources) and sensors entities (sound receivers or listeners in case of a humanoid receiver). Additionally, a geometrical interpretation of the environment is required that is established by a polygonal mesh data format. Such a mesh contains a cloud of three-dimensional points, each labelled by a unique index (vertex). A ring of vertices forms sequences of edges and represents planar surfaces, with the edges forming the boundaries of a finite two-dimensional polygon. The approximation of geometrical structures using such polygonal meshes is not exact and particularly struggles to match curved surfaces. The resolution can be increased by using more polygons, however, a higher number of polygons counteracts the performance of the acoustic propagation algorithms. Therefore, a reasonable compromise must be found to limit the number of polygons as far as possible without introducing perceivable differences in comparison to a more precise resolution.

Entities and surfaces must be enriched by metadata that describe relevant acoustic properties. These entries link functional acoustic models, material datasets or directional databases to provide information on characteristic radiation and attenuation. The acoustic metadata is either interpreted by the simulation engine, or is directly integrated into the audio rendering stage.

4.2.1 Acoustic entities

Acoustic entities are objects that play a special role in a virtual acoustic scene. Such an object features a position and an orientation in a Cartesian coordinate system. It is usually considered moveable and hence can be updated in real-time in a dynamic virtual scene. Although every real object has a spatial extent, in this case, acoustic entities are simplified to infinitely small points in space. Additionally, they inherit specifications depending on their type and purpose. The two most important entities for real-time outdoor auralisation are *source* and *receiver*. Both types are linked to directional functions that describe either the directional character of the emitter or the influences of the receiver. This modification is described by a directional component that integrates the topological shape, fabrics and materials of the spatial object. Consequently, for a valid assumption of a point source and a point receiver, certain distances to the corresponding lotions must be assured.

A *sound source* emits acoustic waves into the propagation medium. The transmitted information can be described as an *acoustic signal* sent out from the source's location with a certain power level. The perceived sound pressure level in space is physically derived from the sound power level L_W of the source. The information emitted by the source and the actual level in the medium can be separated for practical reasons. It is vividly important to have a-priori knowledge on the expected sound power level, because this data is a key feature for prioritisation methods in technical and perceptual acceleration approaches. Further, depending on the type of source, a directivity function or directivity dataset is assigned, for example, to describe energetically the spectral dampening of noise from a car tire taking the vehicle topology into account, which is potentially shadowing strongly to the contra-lateral side.

A *sound receiver* refers to an abstract sensor entity that picks up acoustic information from the medium. Under ideal conditions, this probe object does not interfere with the received sound pressure values. Depending on the type of rendering and reproduction system, it is required to include effects of the human head-and-torso topology and outer ear features, as they play a key role in the spatial perception of sound [Bla97; Vor11]. The Head-Related Transfer Function (HRTF) dataset incorporates these influences and can be interpreted as a *receiver directivity* applied to incoming wavefronts. Attributing binaural filters to a single-channel audio signal are known to reproduce spatial sound convincingly [Bla97]. In this work, a focus on binaural rendering is laid as it appears most appropriate for VR applications and user studies, that are, as a general rule, single-user systems.

4.2.2 Propagation medium

Auralisation applications consider air as propagation medium, because the final receiver is a human in free space. For an urban auralisation application, homogeneous medium conditions regarding static pressure, temperature, humidity and wind speed are assumed. This limitation is essential for the desired GA algorithms, because it allows to fast-forward the propagation simulation between boundaries (cf. Section 3.6). If not otherwise stated, default values for the medium air are chosen according to Table 4.1.

Table 4.1: Default parameters of the homogeneous propagation medium air.

Property	Value	Unit
Static pressure	10125	Pa
Humidity	60	Percent
Temperature	20	Degree centigrade
Wind	[0 0 0]	m/s

The geometrical paths from the pathfinder algorithm deliver the propagation distance value, which is relevant to calculate the attenuation per frequency band. Based on the procedure described in ISO 9613-2 [ISO96], the medium condition and the distance value is used to determine the air absorption attenuation spectrum per individual path during the acoustic modelling.

4.2.3 Geometry format

The outdoor scenario is described by a three-dimensional mesh consisting of points, vertices, *half-edges* and polygons. Polygons are described by a list of half-edges. They are planar and the front face is defined by a normal vector. In particular, half-edges have been chosen over simple edges to exploit the feature of polygon connectivity at those edges. Because an edge is described by two half-edges of opposite direction between two vertices, each half-edge can be linked to an individual polygon. Based on this information, polygons have a mutual connection via the shared edge, which is advantageous for diffraction algorithms. The actual geometrical shape in three-dimensional space is determined by the point cloud, where each vertex is linked to such a point. Half-edge data structures require careful mesh definitions, because a wrongly connected polygonal set can lead to inconsistencies. For example, it is not possible to connect more than two polygons at one edge (e.g., a T structure). Also, the definition of normal vectors is crucial, as it must be consistent to retain a valid half-edge mesh model. For example, if the front side of one polygon of a terrain is flipped, the half-edge structure cannot be prevailed here, as the relevant half-edges do not maintain consistency regarding their directions. In practice this means, that the entire geometry must be generated as a coherent hull that is laid over all objects. This hull then represents pavements, roadways, lawns, buildings, roofs and so on as a whole, and not as separate objects. The decision for a half-edge data format imposes increased demands on the geometrical modelling and requires a careful design process on the one hand, but greatly improves the correctness of the simulation result on the other hand. Structural flaws become apparent during the modelling or the export process and must not be detected during the acoustic simulation or in the auralisation result, where a problem in the model can easily remain undiscovered. Separation of acoustically different surfaces, like concrete walls and soil, is solely realised by the assignment of an acoustic *surface material* per polygon that reflects the impedance of the boundary condition (cf. Section 3.4). As most boundaries found in an urban setting are made of concrete, stone slabs and asphalt, the absorption coefficients are rather low. If no data is linked to the geometrical model, a default value of $\alpha = 0.97$ is selected in accordance to ISO 9613-2 [ISO96].

Because the suggested outdoor propagation simulation algorithm requires pre-processing of the geometry data, a mesh model is considered *static* (cf. Section 3.6). The implementation described in this thesis uses the half-edge mesh model library *OpenMesh*[26].

4.3 Dynamic urban scenes

The most critical feature of a real-time auralisation application is the ability to adapt to a dynamically changing, virtual environment. Naturally, this feature raises great challenges and can only be achieved by formulating limitations that appear disappointing at first glance. To solve a *practical* auralisation problem, it is assumed that a) motion data of a dynamic acoustic entity is not known a-priori, and b) the geometry describing the built environment is static during runtime. Hence, the possibility of pre-calculations is limited to geometry pre-processing

[26] https://www.graphics.rwth-aachen.de/software/openmesh/

and re-use of motion data can only access a history, but is not able to look ahead on a motion trajectory.

A dynamic scene contains moveable objects which are controlled externally by scripting or by motion tracking. With every dynamic object, the computational burden of the auralisation system is increased. Apart from a higher expense on management operations, the two most crucial consequences are laid upon the DSP network (cf. Section 4.7.2) and the simulation back-end workers (cf. Section 4.4). Depending on the scalability and the available hardware resources, feasibility restrictions on the number of dynamic objects are inevitable. According to the guidelines on real-time auralisation expedience formulated in Section 4.5, the DSP network must unconditionally maintain an artefact-free audio rendering, and the simulation scheduler must uphold a sufficient update rate.

An aspect of utmost importance is, that a dynamic scene inherits a dimension of time. The simplicity of this property is most apparent, but the implications on the auralisation system are far-reaching. The classical approach treats an otherwise dynamic scene as temporarily stationary and implements adaption in a frame-by-frame fashion. This method works well for visual rendering, because light travels infinitely fast compared to the temporal resolution of the visual sensory. In contrast, sound travels relatively slow and the claim of simultaneity of an acoustic event with it's perceivable effect must be surrendered. The idea to react upon scene modifications by sequentially executing acoustic simulations to provide (steady-state) IRs and update DSP (convolution) elements at a sufficiently high refresh rate is only valid, if the dynamic component can be considered quasi-stationary for each time frame (slow motion). This assumption is violated by fast-moving sound entities.

A divestment is suggested and justified in order to replace the classical method by a heuristic time concept for a dynamic acoustic scene. A timeline becomes necessary when designing a comprehensive auralisation framework that thoroughly integrates the Doppler shift.

4.3.1 The auralisation timeline

To describe spatio-temporal changes in a dynamic urban scenario, a timeline concept is introduced that attaches a time stamp to any event related to motion. In particular, initial source and receiver motion is recorded and buffered for a sufficiently long time period for direct use in specialised units of the rendering

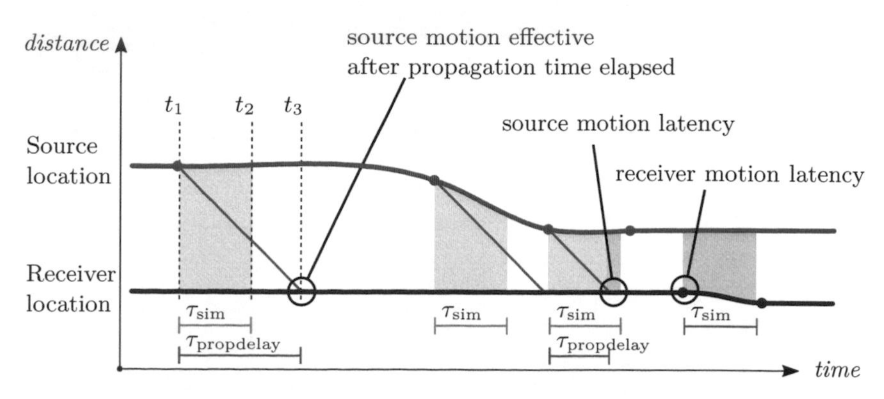

Figure 4.2: Auralisation timeline example.

module. Additionally, simulation tasks are assigned a creation time stamp and corresponding results are sorted into chronological series. This theoretical concept makes it possible to render dynamic events within the DSP chain at the moment they become audible with respect to the receiver location. The approach is only useful if scene data (motion, simulation) is evaluated on different levels in the rendering pipeline *and* simulation results deliver, to a certain degree, separable propagation features per individual geometrical path[27]. In other words, the rendering stage does not entirely rely on simulated data (which has a high probability to be flawed by latency), but also bases rudimentary update routines on initial scene information, such as raw motion streams. The suggested rendering model for dynamic urban environments requires an interpretation of sound propagation as individually travelling wavefronts. A simulation result is subdivided into propagation values of a delay, attenuation spectra and directions regarding the source and receiver location and orientation (cf. Table 4.2).

Figure 4.2 depicts a source-receiver trajectory example under static free-field conditions that illustrates the correlation of separate translation events on the simulation trigger and the chronology of results. τ_{sim} describes the constant simulation time and $\tau_{\mathrm{propdelay}}$ the propagation delay at the receiver location regarding the past source location (at corresponding retarded time, see [Str98]).

The initial source motion triggers a propagation simulation at time stamp t_1 for the current receiver location, which becomes available after τ_{sim} elapses at t_2 and

[27] The very opposite of this feature is an auralisation approach using Binaural Impulse Responses *(BIRs)*, which integrates every propagation aspect into a single representation – including receiver orientation.

is inserted into the result history at creation time stamp t_1. Because the source motion does not become effective at the receiver location *before* the propagation delay $\tau_{\text{propdelay}}$ has passed, the auralisation operates without latency induced by the propagation simulation. As soon as the source's distance to the receiver undergoes a certain limit, the simulation time exceeds the propagation delay, as visualised by the third motion step in the figure. If this happens, the renderer can only rely on outdated result items in the history, which produces latency. In this case, one must either accept a system latency at a magnitude of the simulation time τ_{sim} in the worst case, or the source's initial motion stream must be used to estimate the corresponding acoustic effect, for example, by an extrapolation model. An even more severe problem becomes apparent for receiver motion, reflected by the light-red block in the figure. The coincident timing of trigger and result basically implies that the simulation time must approach zero, which is impossible. However, a perceivable effect must be performed *instantly* to produce a convincing real-time auralisation response. Again, a solution that circumvents an added latency of τ_{sim} is proposed by using the initial receiver motion values in order to apply a predicted perceptual response and take measures to provide estimated coefficients for the DSP network.

This concept is able to deal with two kinds of challenges that are of particular importance in dynamic urban environments. Firstly, a source motion history allows to employ simple interpolation and smoothing routines to produce sampled, synchronous trajectories. This option is extremely important to auralise the source-related Doppler shift without artefacts. Secondly, receiver motion in general, and receiver rotation in particular, can be directly integrated into rudimentary receiver spatialisation processing units that do not require a full propagation simulation (i.e., are robust enough to be merged with out-of-date simulation results with a convincing approximated result). Especially the immediate response to user head rotation must be outlined, as it can be implemented with minimal latency. Considering a binaural approach using convolution of HRTFs, head rotation triggers a filter coefficient update adapting the new local directions of incidence. If the spatialisation unit directly applies these updates based on the *initial* latency-free motion stream with respect to potentially outdated incident wavefronts from the simulation history, a perceptually convincing auralisation output is presented without added latency from the simulation. However, this method requires a per-path incident direction handling, which is only feasible with a binaural directional clustering algorithm, as will be presented in Section 4.7.2.

4.3.2 Motion input

In practice, motion in a dynamic virtual environment is provided by scripting and by motion tracking systems. For simplification reasons, it is assumed that sound sources are controlled by scripting and a user is linked to a virtual receiver controlled by a stream of motion data from a tracking device (sending position and orientation at a reliably constant update rate). However, the roles can be switched without restriction. An artefact-free auralisation of an externally provided, unprocessed motion trajectory is likely to fail, because the rates at which a rendering system demands coefficient updates are magnitudes apart (e.g., 30 Hz motion frame rate in a VR environment vs. 800 Hz block rate in the audio device). Consequently, enhancements must be applied by the auralisation system to improve the quality of incoming trajectories and up-sample the DSP coefficients.

Trajectory scripting

Moving sound sources following a sampled trajectory are a common auralisation use case. A straightforward solution is based on scripting. The realisation commonly provides locations of a given spatial and temporal resolution to update the virtual scene step-by-step in the expectation to auralise, for example, a continuous vehicle pass-by event at a constant velocity. The motion input stream can be flawed regarding the *timing*, which must be addressed. In addition, the spatial sampling resolution is seldom considered by the user leading to non-conformity with the sampling theorem, which makes a correct reconstruction of the intended trajectory impossible. However, an auralisation framework must be able to handle scripted trajectory data by providing methods to deal with potentially under-sampled motion data. Two methods are proposed: *low-pass filtering* and *timer synchronisation*.

To improve the trajectory input in a user-friendly fashion, it is suggested to treat any incoming motion data as a noisy signal stream of three-dimensional spatial data and apply an adequate improvement for the auralisation engine. An appropriate smoothing routine with a low-pass character is applied to the incoming data and a re-sampling of the output is performed. By doing so, an alias-free trajectory can be produced that is in synchronisation with the audio processor. For example, a reasonable model to improve the quality is applying a physical inertia force that is assigned to the sound source ensuring that the temporal derivates of the trajectory are continuous. This approach effectively generates motion of continuously differentiable velocity, acceleration or even jerk (the latter gives the best result for a seamless Doppler shift) at the expense of interpolated motion that can deviate substantially from the initial trajectory at critical locations. For example, if motion input for a virtual object with a great mass is provided that includes a steep, right-angled segment, the inertia principle will force the trajectory to pass this point in the initial direction and slowly approach the new direction after a given time. The resulting curve is continuous and provides a plausible auralisation yet differs from the desired path.

Considering motion data with a flawed temporal synchronisation, it is proposed to apply non-linear smoothing routines that can handle non-equivalently sampled signals, like a moving average filter or a Lagrange interpolation. However, if trajectory smoothing is not desired by the user a method must be provided to ensure synchronisation with the auralisation timeline. A feasible option is the combination of a locking mechanism that collects scene changes and handles them as an atomic modification with a high-precision blocking timer to synchronise client update loops. Scripting timeout-based update loops iterating over samples of a trajectory with the intention to produce an *isochronous* update mechanism is often found. However, this solution causes temporal variations that result in a perceivable Doppler modulation. The problem lies in the insufficient function sets of client scripting workspaces. In addition, function calls evoked within a loop body are subject to processing time deviations, which can be unpleasantly high for an auralisation system. From a client's perspective, irreproducible results are obtained that appear as random auralisation artefacts, especially, if the Doppler shift is not correctly applied due to re-sampling issues of flawed trajectories.

Listing 4.1: Sampled trajectory, linear pass-by motion (MATLAB® code).

```
1  %% Trajectory setup
2  f_s = 100; % Hz update rate
3  v = 20; % 20 m/s velocity
4  d = 200; % 200 m total trajectory distance
5  N = d / v * f_s; % Number of samples = 1000
6  x = linspace( -1, 1, N ) * d / 2; % x = -100m ... 100m
7  set_auralisation_timer( 1 / f_s ) % High-precision timer
8
9  %% Update loop
10 for n = 1:N
11
12     % Locked scene modification
13     lock_update()
14     set_sound_source_position( [ x( n ) 0 0 ] );
15
16     % ... more time-synchronous modifications
```

```
17
18        % Timeout
19        % pause( 1 / f_s ) % (not recommended)
20        wait_for_auralisation_timer() % (blocking wait)
21
22        unlock_update() % transmit isochronous updates
23
24    end
```

Listing 4.1 shows a code example for a smooth pass-by auralisation taken from the documentation of Virtual Acoustics *(VA)*. A high-precision timer is pre-configured for an initial timeout providing the desired update rate (line 7) and a blocking wait function within the update loop is executed in order to produce isochronous motion (lines 20 and 22). Line 19 shows the alternative time-out mechanism provided by the scripting environment, which is not recommended as it is not sufficiently precise. In addition, a locking mechanism for scene updates is used, which groups motion updates in order to indicate that all transmitted calls (substituting line 16) shall be treated as a single scene modification with the same time stamp (e.g., if a virtual sound object is represented by a cluster of single point source).

Motion tracking

Motion tracking systems are expected to observe the rotation and translation of an object with a sufficiently high and time-synchronous frame rate. Furthermore, it is assumed that a decent tracking device provides a motion stream that complies with the sampling theorem making complex refinement obsolete. However, if either the tracker frame delivery is asynchronous or an intolerable imbalance between motion update rate and audio block rate of the rendering thread is observed, the above-mentioned methods apply (at the expense of added latency). Typically, a light motion smoothing supports the elimination of unwanted artefacts in the Doppler shift simulation.

Timers and clocks

On a technical note, because a computer system can have many different clocks and various parallel threads, it is important to settle for one that is accessible in the global scope and is highly accurate. A high-performance timer provided by the operating system is sufficient, however, with a quality audio interface using an internal hardware-based clock and a low-latency audio frame configuration, it is reasonable to rely on the streaming timer of the audio context.

4.3.3 Data history and parameter interpolation

The concept of an auralisation timeline introduced in Section 4.3.1 is helpful to chronologically align input data, for example, motion of sources and receivers. The timeline also contains simulation results delivered by the simulation scheduler (result history), which will be presented in Section 4.4.2. Every result item refers to a simulation request (task) from a time snapshot in the past, that is approximately outdated by the simulation calculation time τ_{sim} and interprets the scenario as momentarily stationary (cf. Figure 4.2). Because the auralisation timeline regards the present time from the receiver perspective, events concerning the source lie in the past under the condition that movements are not exceeding the speed of sound. To make source-related sound propagation effects audible at the right point in time, it is necessary to imagine that every single wavefront arriving at the receiver location corresponds to an individual geometrical propagation path. Each path incorporates a propagation delay pointing back in time to a past source location. However, if the simulation time is not infinitesimally small, this concept develops an implicit dilemma. Theoretically, it is impossible to determine the retarded source location without knowledge on the individual, per-path propagation delay from the simulation, which requires the current receiver location to finally determine the retarded time itself. Analytically, the following recursive term must be solved in order to determine the propagation delay correctly (exemplary for the free path):

$$\tau(t) = \frac{\|p_r(t) - p_s(t - \tau(t))\|}{c}$$

with t the current time, τ the propagation time, c the speed of sound (homogeneous medium) p_s and p_r the source and receiver location, respectively. This function is continuous and bidirectional, as long as the speed of sound limit is not exceeded. To solve this problem, the strict temporal requirement of receiver location and propagation time, as issued by the formula, is abandoned and an added latency from the simulation determining the propagation delay must be accepted.

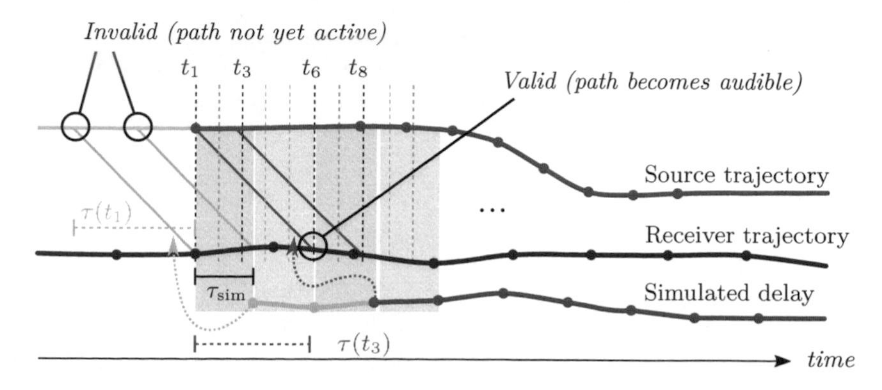

Figure 4.3: Determining the propagation delay with the auralisation timeline.

In practice, instead of solving the analytical formula per propagation path (including image sources etc.), a time scanning and locking mechanism for the individual propagation delay can be implemented, that exploits the continuous, bidirectional property of the propagation delay function over time. Figure 4.3 visualises the procedure by qualitatively plotting a source and receiver trajectory as well as the corresponding simulated delay for a geometrical path (potentially including reflections and diffractions). The audio frame rate of the system is indicated by vertical, dashed lines, and both source and receiver data is up-sampled to the audio frame rate (indicated by continuous curves) as they are potentially non-isochronous and relatively sparse (cf. Section 4.3.1 and Section 4.3.2). To determine the acoustically relevant contribution of the *retarded* source (i.e., signal, directivity and Doppler shift from the past) via the given path, the latest simulation result is consulted and the delivered propagation delay is evaluated. Then, the data history of the sound source is queried at the corresponding date (current time minus propagation delay). If no data is present, the corresponding path is considered as not yet audible. In Figure 4.3, a simulation is triggered at time t_1 (source is created) and the result is available after τ_{sim} elapses (green area), when another simulation is triggered (light green area) and so on.[28] The resulting series of propagation delay values are represented by the third purple curve with dots where samples are inserted beginning with arrival of the first simulation result from frame one holding value $\tau(t_1)$. Scanning the history of the source data looking into the past by a period of $\tau(t_1)$ returns an invalid range for the first frame. Accordingly, the second simulation that is using the latest available scene snapshot at time t_3 (frame 3) delivers a non-valid source

[28] This routine assumes sequential simulations of constant runtime for simplification reasons.

state. The subsequent simulation based on frame 6 at time t_1, delivers a first valid source state from the source's data history. At this point, a sound source assumed stationary in location becomes audible to the receiver, because the elapsed propagation time $\tau(t_3)$ of frame 3 is based on source and receiver locations evaluated at time t_3. However, the source may have moved prior to t_3 from a location at t_1 (a period influenced by the simulation runtime τ_{sim}). To finally solve this issue, the approach must re-iterate the simulation using the current receiver location at t and the latest retarded source location at $t - \tau(t')$, until a sufficient resolution for the retarded source location is found.

An iterative propagation simulation to resolve this temporal misalignment poses a high computational burden and produces latency at multiples of the simulation runtime. Instead, an informed interpretation of the available simulation results from the result history is suggested (cf. purple sequence in Figure 4.3). The imposed deviation from the correct timeline is only severe for fast moving sound entities due to the false assumption of a stationary scene in order to determine the propagation delay. Based on the motion data from the history, it is possible to derive a velocity confidence value separately for the source and receiver. This value can be used to optimally parametrise an interpolation algorithm for the simulation results that evaluates and weighs the propagation features, namely the propagation attenuation spectrum, delivered by two simulation results: the first result that corresponds to the propagation delay with a falsely-timed source location, and the second result that is chronologically closest to the current time minus the propagation delay, which is based on a receiver location from the past. The basic principle of this propagation parameter interpolation approach is, that a good estimation of the temporally congruent simulation result can be made, if one of the entities is not moving fast. In particular, this is the case in VR applications with a user that is only moving in a limited area by own body motion.

4.4 Propagation simulation

The *simulation context* is a control thread entrusted with the transformation of a virtual environment into parameters that reflect the propagation of sound of a scenario. A physically motivated simulation must be performed, which considers the environment to be momentarily stationary. Results must be interpreted, transformed and delivered to the audio context where the data is needed to manage and control a DSP network, as will be introduced in Section 4.7.2.

The general employment of propagation simulation for auralisation purposes is rather simple: considering a sequential processing, a scene modification triggers a simulation and an intermediate result is made available. This intermediate result is then transformed into DSP parameters, for instance, an IR for a convolution unit, and the corresponding processing components in the rendering pipeline are updated. From this point on, the auralisation output delivers an audio stream that represents the latest scene state, until another update is delivered. In general and for maximum auralisation quality, the steps are performed frame-by-frame producing a short audio output sequence per simulation (e.g., several hundred samples per block). However, the computation time of this approach exceeds the resulting auralisation track length. Therefore, it is only practicable for offline rendering, as the real-time constraint is violated.

In contrast, in a real-time auralisation system, the audio context and the simulation context must be decoupled and separately executed in two parallel threads. Furthermore, to avoid a core system overload, it is wise to separate the auralisation unit from the simulation execution and distribute the (heavy-duty) simulation execution by means of low-priority threads – or even outsource the calculations to another machine [Wef+14]. The implementation of this concept requires a *scheduler* that accepts tasks and assigns them to back-end worker nodes. Every task must contain all current scene information, for instance, the locations and orientations of sources and receivers.[29] If a free worker is assigned a task, the node's simulation instance extracts the task information and performs an acoustic propagation simulation, for example, based on the pathfinder algorithm from Section 3.6, followed by an acoustic propagation modelling, as discussed in Section 3.4. When the result is available, a container is assembled with the requested propagation parameters and returned to the client (e.g., a list of paths with dedicated DSP parameters for the rendering stage, see Table 4.2).

To integrate scheduling, simulation tasks must be triggered from the audio context at the audio frame rate based on locations obtained from the interpolated and up-sampled motion data of sources and receivers. This is important, because entity trajectories are smoothed and parameters for the DSP network rely on interpolated information. If simulations are instead triggered by sporadic asynchronous scene updates, a problematic divergence in the auralisation timeline occurs. If intermediate values based on the up-sampled approximated motion stream are mixed with raw user locations, fluctuations, modulations and dis-

[29] Because the geometry of the virtual environment is considered *static*, further scene information is not required, as each simulation worker loads the scene data in the initialisation phase.

continuities are likely to appear in the stream of DSP parameters. The first consequence is, that a small time budget must be assigned in the highly critical audio process to assemble tasks and propagate them to the scheduler instance. However, with real-time processing methods, the footprint is minimal and is justified (cf. Section 4.5.4). A second issue is identified by an extensive input stream of tasks issued to the scheduler, which has to be managed. Compared to practical propagation simulation times, the generation of tasks for each audio frame is insufficient, as only a fraction can be simulated while the majority of tasks must be rejected. However, a performant scheduling strategy is proposed that effectively prevents a drain of precious resources (cf. Section 4.4.2).

4.4.1 Integration of propagation path results

A successful propagation simulation returns a list of individual propagation paths for each source-receiver-pair consisting of specific propagation parameters that are required by the auralisation rendering stage. The result item is tagged with the task creation time stamp and corresponds to a quasi-stationary acoustic interpretation of the virtual environment at the corresponding time that is roughly outdated by the simulation calculation time. The result is not directly included into the rendering process, because some parameters require an interpolation with history data (cf. Section 4.3.3). The list of parameters needed by the rendering approach according to Section 4.7 is shown in Table 4.2.

Table 4.2: Set of parameters describing an individual propagation path.

Parameter	Purpose
time_stamp	Task creation time stamp
identifier	Used to relate the result to a path in the rendering
propagation_delay	Propagation delay of the geometrical path
spreading_loss	Spreading loss gain
geo_attenuation	Geometry attenuation spectrum (reflection, diffraction)
air_attenuation	Air absorption attenuation spectrum
incoming_dir	Normalised direction of incoming wavefront at receiver
outgoing_dir	Normalised direction of outgoing wavefront at source

4.4.2 Scheduling

If the simulation of sound propagation is resource heavy and the issued update requests exceed the number of results that can be produced, a scheduling system is advised. A scheduler receives tasks from a client and distributes them to a number of worker nodes following a strategy. Worker nodes are running in the background and instantiate a sound propagation simulation engine that is exclusively used to deliver one result per assigned task (i.e., one task for one source-receiver pair). Workers are scalable in number and ideally have access to exclusive resources for an increased performance, for instance, by executing tasks on distributed machines like a high-performance computing cluster [Wef+14]. Results provided by worker nodes are handed back to the scheduling instance and are forwarded to the client. The quality of a scheduling system is measured by the throughput capacity and the scaling ability.

Introducing a scheduling unit to an auralisation application increases it's structural complexity on the one hand, but significantly improves the scalability and universality on the other. However, if simulations become as complex as in urban environments and additionally must be executed *in parallel* for many source-receiver pairs, the integration of a scheduling approach is inevitable. A concept is proposed that uses a lightweight and universal interface designed to manage an extensive amount of incoming tasks and a sophisticated performant filter routine that rejects tasks on two conditions. Firstly, only updates are considered that pass a user-defined *relevance test* evaluating a logic decision network of conditional blocks. Secondly, a task initially marked relevant is omitted if the simulation back-end is overloaded and an equal task with current scene data is issued. In this case, a displacement algorithm is applied and the outdated task is replaced by the recent task in the queue.

Logic decision network

Strategies to realise a logic network deciding that an incoming task is more relevant than the queued tasks can be versatile. Since an appropriate scheduling strategy is highly context-dependent, a user-configurable logic decision network has been implemented with a function that compares two tasks. Some fundamental tools and building blocks are required to design a decision network, namely a logical AND connection, a logical OR connection and an inverter option (NOT function). A logical unit is a specialisation of an abstract base class that implements a comparative operation of selective parameters of two tasks that returns a positive or negative result. The combination of such logical units to a network of parallel and sequential routes decides, if a new tasks replaces an old task. It is imperative that the comparison is performed rapidly and with minimal computational effort, because it is applied to every single incoming task. For auralisation purposes, decision units are performing an *importance sampling* that can be separated into classes of *technical* and *perceptual* methods.

Technical importance sampling

The technical class aggregates non-acoustic mechanisms and physically motivated considerations that are based on known acoustic effects.

Measures to filter incoming tasks for technical reasons are helpful to design a robust and efficient scheduling system. For instance, a timeout-based filtering is a simple way to reduce an incoming stream of tasks, if the initial task stream rate is not required. If the simulation back-end has a relatively long mean computation time that is out of proportion to the task stream, a timeout-based filter is a legitimate measure to rapidly purge tasks at basically no cost.[30]

Another helpful and straightforward idea to cut down potentially similar update requests without considering the acoustic impact are distance and angular threshold filters. An incoming update is only marked important, if a self-related, geometrical metric is exceeded (e.g., a source has moved at least 10 cm) or a relative distance is modified significantly. The reasoning behind this concept is, that a movement as small as millimetres is acoustically irrelevant. However, in practice, motion tracking values can be noisy and trigger an update per data frame propagating the noise to the auralisation motion timeline. These kind of system-related specifics are effectively suppressed in the scheduler with technical threshold filters that are unlikely to have an impact on the auralisation result, if conservatively adjusted.

The options to realise logical units are countless and can even incorporate scene information, if applicable, and is the main reason why the decision network is realised to be configurable by the system designer.

[30] The only draw-back conceivable is an additional latency between issued task and delivered result, which in worst case is in the range of the user-defined timeout – that is usually irrelevant w.r.t. the simulation time.

Perceptual importance sampling

The idea to include perceptual aspects into the importance decision process of an update has been raised during the implementation of an urban auralisation framework. The magnitude of sound sources and resulting propagation paths leads to an overflow of simulations if an otherwise uninformed update handling is employed. The concept to consider perceptually relevant updates is evident, however, requires to interpret the virtual scene data psycho-acoustically, which is generally the duty of an analysis module that must interpret simulation results and is potentially costly. However, it is highly relevant to pursue the realisation of a detection algorithm for dynamic scenes that filters perceptually important updates. Consequently, perceptual importance sampling implies predictive models. Several acoustic features can be evaluated relatively fast and are conclusive for the perceptual importance sampling without dedicated experimental validation, if carefully and conservatively applied. For example, rotation movements of source and receiver entities are only important, if perceivable angular thresholds are exceeded [FZ07; Bla97]. To give an example, if a-priori knowledge on the source's sound power is available, the source location can be used to define a distance metric representing a region-of-interest that can be compared against other source in order to generate an importance ranking [CMS08]. These kinds of criteria, however, are highly context-dependent and require an individual verification. Consequently, if concepts like temporal and spectral masking of the human's auditory system are desired to be incorporated in the decision tree, a corresponding masker signal must be present. Assumptions are conceivable for certain constellations, for example, if a stationary wide-band masker (e.g., a water fountain) is present at all times within a restricted area.

Adaptive importance sampling

The imbalance of propagation paths found in a multi-source urban environment and the actually processable amount at real-time rates is identified as the major issues of the presented approach, as will be addressed in Section 5.5. A great potential may be scooped from *adaptive importance sampling* routines for automated urban auralisation applications in the future that considers the maximum processable path count in the decision process. Adaptive scheduling algorithms must quickly interpret the current situation acoustically and derive importance ranking on perceptual aspects for the most important paths. Basically, the question is raised, which sources and corresponding propagation paths must be included in the auralisation rendering pipeline considering all other sources that contribute to the overall soundscape, indicating a recursive problem. The integration of adaptive importance sampling integrated into the logic decision network of the simulation scheduler is a promising idea to counter the computational capacity shortcoming that inevitably occurs with increasing complexity of virtual multi-source environments, for which in turn psycho-acoustic masking effects gain relevance. However, it appears challenging to obtain robust reasonable statements of universal applicability, as the implications are highly context-dependent. Foremost, the entanglement of predictive measures based on acoustic considerations alongside the lack of actual simulation results at that stage, as aforementioned, makes a conscientious perceptual validation by the system designer necessary. This topic is discussed in Chapter 6 and in the outlook of Chapter 7.

4.5 Real-time processing

The basic principle of a program operating under real-time constraints is simple: upon parameter change, a system's reaction must be executed within the allocated time budget using the available resources. If this principle is violated, the system is not operating correctly. However, the implications are far-reaching and are briefly discussed in the context of outdoor acoustics.

Performing auralisation under real-time constraints is challenging in many aspects. Because the application of real-time auralisation is diverse, it is indispensable to exploit all acceptable means to break down requirements to an absolute minimum. To do so, it is vital to sketch the use case and derive adequate limitations without interfering with the purpose of the application. Two major axes that drive the decision process can be drawn: one that considers the objective of the scenario and one that bears relevant perceptual aspects in mind. In general, a sufficient real-time auralisation must operate artefact-free in the first place, meaning that there will be no audible dropouts, clicks, noises or other unwanted distortions in the output when running the program for the designed purpose. Of equal importance is, that the auralisation reactivity complies with relevant perceptual thresholds. For example, update rates must be delivered at sufficiently high rates and simulations must maintain an acceptable input-to-output latency.

4.5.1 Technical aspects

In many cases, if real-time processing is desired, exterior technical circumstances like the available computational power dictate limitations regarding the extent of the virtual environment and number of entities. Although the actual implications are varying, the run-times usually scale with

- the number of sound sources,
- the number of sound receivers and
- the number of geometrical polygons.

If the employed algorithms do not automatically restrict these numbers, the designer must ensure a manageable quantity, for instance, by limiting the geometry mesh resolution and the number of acoustic entities (cf. Section 4.2.1).

Regarding outdoor noise assessment applications, the most challenging task is to spatially render many dynamic sound sources in real-time. The *fast* simulation of sound propagation inside the urban environment is a hard problem and, in case of a deterministic geometrical approach that does not apply sophisticated acceleration methods, scales exponentially with the number of polygons. Distributed computing for polygon-rich urban models separating tasks per acoustic entity is one solution to face computational resource shortage (cf. Section 2.6, see also [Wef+14])). The rendering of individual time-variant propagation paths for dynamic sound sources consumes significant processing power in a prioritised real-time processing thread, namely the audio context (cf. Section 4.7.1). In this context, the available time budget for calculations is extremely limited. The strict timing of the audio callback function driving the auralisation stream is incompatible with a distribution concept, as any delayed transfer of data leads to audible dropouts in the output leaving only tailored specialisations or even dedicated hardware features a feasible alternative. Since in a VR application the auralisation is often performed in parallel to the visual rendering on the same machine, the distribution of processing power to the graphics hardware (GPU) has not been pursued. The audio processing is solely considered to use the CPU of a PC and implications on the auralisation capacity are accepted. Furthermore, demands are made that the realisation of an auralisation system operating under real-time constraints must realise a certain set of relevant acoustic phenomena. The following per-path aspects are considered non-negligible for outdoor acoustics:

- Frequency-dependent propagation attenuation (absorption, diffraction, directivity, etc.),
- The Doppler shift caused by entities moving relative to the propagation medium and
- Localisation features of incoming wavefronts to create a spatial acoustic impression at the receiver location.

Based on this catalogue of requirements, a selection of DSP units must be employed and a real-time capable network structure based on those units reflecting the virtual acoustic environment is proposed. Because the processing network is part of the audio context, it consumes valuable resources of the time-critical audio streaming thread. This imposes an extremely effective design and requires that every single unit of the processing network must be realised with the highest possible efficiency. Applicable DSP units are outlined in Section 4.6. An example configuration is presented in Section 5.5 and the feasibility for virtual environments is discussed in Chapter 6.

4.5.2 Perceptual aspects

By all means, auralisation aims at creating an experience that is perceived by the human auditory system. For this reason, it is permitted to economise on every part of the system, as long as the implications are impossible to be detected by the user. Viable possibilities are as diverse as the human auditory system's capacities, ranging from fundamental physiological limitations to cognitive processing of sound information. From an implementation viewpoint, this circumstance is both a curse and a blessing. On the one hand, it creates countless prospects to cut down memory consumption and computation time, which is appealing to any developer. On the other hand, it can be a difficult task to reliably anticipate the events of an interactive virtual environment and take advantage of withdrawing potentially imperceivable calculations given the complex processes of sound perception and cognition. Further, to provide conclusive proof of a perception-related simplification, an investigation based on listening experiments must be performed, which can be expensive and time-consuming as it often serves one purpose only. To give a simple example, the selective attention ability is an attractive aspect to be exploited in the context of auralisation. A liable prediction could steer the resource distribution. Also, if sound is accompanied by the visual modality and a user is assigned a certain task in a virtual reality environment, the user's occupation on the visualisation and task performance broadly withdraws attention from the acoustic perception, leaving room to lower the demands on auralisation in both quality and quantity. These (and potentially many more) elaborated perceptual acceleration aspects of real-time auralisation in the context of outdoor environments are subject to future research, see Chapter 7.

Some rather general and scientifically sound aspects of perception have been incorporated, because they are crucial to approach real-time update rates and low-latency processing. Matching the perceivable frequency range is one example that is a universal reason to select an appropriate audio system sampling rate of 44.1 kHz. Also, without formal proof by a listening experiment, the atomic latency unit is chosen to a value of 1.2 ms, a time frame resolution that corresponds to an audio update rate of 833 Hz (cf. Section 4.5.2 and Section 4.7.1). This relatively challenging latency setup has been found to deliver a sufficiently reactive audio rendering under free-field conditions. However, it is likely that this conservatively chosen configuration can be significantly lowered in more reverberant spaces.

It is formally refrained from any further discourse on perceptual acceleration techniques as no claim of general validity can be substantiated without conduction of listening experiments. The explicit validity of employment and the selection of parameters of technical and perceptual acceleration features lies within the responsibilities of the system designer, as exemplary put into praxis by Ehret et al. [Ehr+20]. This is not otherwise conceivable, as the virtual environment and the appropriate configuration of the auralisation framework is entirely context dependent and scales with the available resources and tangible recommendations are outside the scope of this thesis.

Instead, a system example is discussed that alleges conformity to the basic principle of real-time processing, namely artefact-free rendering undergoing perceptual thresholds (cf. Section 4.7). A demonstration of what is achievable is reviewed with respect to a generic outdoor auralisation application on a current PC, which assumes that the propagation effects discussed in Section 4.5.2 are to be incorporated and a binaural output is desired. It is anticipated to demonstrate the methodology that realises the real-time auralisation of a rudimentary outdoor scenario in a realistic fashion without exploiting every possibility to maximise, for example, the number of paths that can be processed for a certain configuration and hardware setup. The impact of perception-based control measures on real-time capacities is selected in a conservative way leaving generous potential for improvement, which in turn must be proven valid in the context of application. Sizeable acceleration results can be achieved in the outdoor simulation algorithms of sound propagation (cf. Section 3.6 and Section 4.4), leading to higher parameter update rates. With a binaural clustering method for incident wavefronts and a reasonable configuration of dimension and quality of DSP units, the computational load is broadly lowered in the audio rendering process leading to an acceptable overall performance for demonstration purposes (cf. Section 5.5).

4.5.3 Latency

In dynamic virtual acoustic environments, *latency* is considered as the time discrepancy between the response of the technical system to a user input (or any other state change) compared to the equivalent real-world situation [Wen98]. The auralisation of a dynamic virtual outdoor scenario generally requires low-latency parameter adjustment whenever a sound source moves in order to generate the required immersion. Of equal importance is, that the system adapts to user motion under real-time constraints considering head rotation *and* body translation. Put simple, if the user turns, the location of incident directional sound must remain stable considering the world coordinate system, which requires rapid DSP coefficient updates. Furthermore, propagation effects must be recalculated fast enough to cover the expected effects with approximate accuracy concerning the timing. For example, if a traffic vehicle disappears behind a noise barrier or building, a dampening effect of the source's sound signal must be audible at the expected point in time. However, the demands on latency thresholds depend on perception and cognition capabilities. Environments with distinguishable directional sound field components may be more stringent on latency values than diffuse environments with long reverberation tails [Vor11]. For real-time auralisation systems, as a rule of thumb, the requirement to adapt to a state change is highest for receiver motion and decreases with distance from the receiver, because the propagation delay can be exploited. Consequently, it is particularly important to ensure low-latency updates for user head *rotation*, as this interaction is expected to trigger an immediate audible response. If the directional processing of a sound field at the receiver is slightly delayed, the auralisation system appears inert and immersion is reduced. Fortunately, if far-field assumptions hold, as is the required by GA algorithms, a receiver rotation does not require a new sound propagation simulation. It only affects the spatialisation component that handles the incident directional sound calculation. Considering different approaches to render and reproduce sound, approaches like HOA and VBAP are by design maintaining the freedom of head rotation and are therefore predestined for real-time applications with a fixed receiver space. Binaural algorithms require a re-interpretation of incident wavefronts and must perform a low-latency filter exchange of the corresponding directions by querying the HRTF dataset.

Of equal importance is the adaption to translation of acoustic entities in a virtual environment. These kind of state changes are challenging in terms of latency, because a re-iteration of the sound propagation calculation is required. In theory, a full simulation cycle plus parameter exchange in the audio process must be executed, which can introduce substantial delay. Firstly, the impact of movement on the perceived sound at the receiver is decreasing with distance among entities and boundaries considering the acoustic models outlined in Section 3.4. For instance, a relatively small sound field modification can be expected, if far-off sound source approaches the receiver by a fraction of the distance. Hence, higher latency values are acceptable, as long as perceptual implications are modest. In turn, a sound source that passes a gap of a noise barrier raises higher demands on the latency criterion. According to diffraction models, attenuation values vary largely for small distances in this constellation (cf. Section 3.5). Because the effect is assumed to be clearly audible, a timing mismatch of the auralisation caused by latency is more obvious and therefore easier to identify. Secondly, a source movement may be visible immediately in a real situation but the audible effect is not effective at the receiver location until the propagation time elapses. This delay value can be subtracted in the latency equation for sound source movements. However, it must be emphasised that this does *not* apply for receiver movements, as this interaction would trigger an immediate response (similar to receiver head rotation). Again, if a method is robust against receiver movements within a certain area, recalculations on the propagation part can be skipped to a certain extent making these approaches predestined for applications with limited receiver movement.

It is assumed in the context of outdoor auralisation for VR applications, that sound source and receiver entities are able to move freely and without predetermination in three-dimensional space, while the built environment of the urban setting is considered static. The circumstance of unpredictable movement basically neglects propagation pre-calculations and forces to apply real-time algorithms that must execute sound propagation simulation and the transformation into DSP parameters for audio processing with low latency.

4.5.4 Parallel programming concepts

In computer programs with real-time processes, *threads* create procedural contexts that are executed in parallel. Threads are an integral part of real-time auralisation applications implemented in software for an architecture using a multi-core CPU. Employing threads enables to design a dynamic framework that handles separate modules independently, while the operating system can assign computation capacity according to a scheduling principle that also takes the priority of a thread into account. The audio processing thread receives the highest priority and is synchronously triggered by the audio hardware at relatively short time intervals. A rendering module is part of this thread and produces output samples by executing per-frame calculations based on a network of Digital Signal Processing (DSP) units (operators, delays, convolutions, etc., cf. Section 4.6) on an input stream (i.e., a signal source). These units are modified by parameter exchange methods that are called by another thread and smoothly blend to produce a continuous artefact-free output. As long as there are no further updates delivered, the audio thread will continue to process the input stream as if the virtual acoustic environment is stationary. Providing an update to the DSP units immediately produces a modification in the output stream and the processing continues, until another update is triggered. This way, a continuous output signal is generated and an update routine that is independent of the audio thread delivers scene changes. These scene changes are not necessarily synchronous and are usually triggered by user motion from a tracking device or by push messages from a simulation scheduler delivering an update from the sound propagation calculations.

A delicate problem considering threads in general and the audio thread in particular is the proper exchange of data. Although both threads share the same memory, the program must ensure that the concurrent access to the data is guaranteed without interfering the audio processing routine. In other words, an update procedure cannot directly override DSP coefficient memory in the audio context, as a data read process may be accessing it contemporaneously, leading to a mixture of old and new values with unpredictable outcome. To achieve a consistent handshake during data exchange, either atomic data formats must be used, or non-blocking data synchronisation methods must be applied. Atomic data formats inherently perform an operation such as a read or write procedure without the possibility to be interrupted. They are conceivable for single values such as the overall gain of an amplification DSP unit and are robust against concurrent read/write access. However, the atomic concept does not offer to decide which process is suspended on access and potentially develops a side effect on the audio processing budget. Consequently, is insufficient for large data fields, like hundreds of filter coefficients.

Therefore, the exchange of data beyond single values must implement a non-blocking concept using mutex or critical sections. Both methods offer the possibility to lock and unlock execution blocks and also have the option to inquire, if such a block is currently locked. This way, a DSP unit of the audio context can provide a secure update method that mediates the incoming data by buffering it first while potentially suspending the sending routine, and swap the data at a safe point in the time-critical routine at it's discretion, for example, at the beginning of each processing block. This way, the audio process is not blocked by an outside call and the concurrent data access is minimal. Non-blocking data exchange is ideal for rarely occurring updates, for example, if DSP coefficients from a simulation result are transmitted.

A similar concept can be realised with a *concurrent queue*, if an excessive write access is assumed. Queuing with an extremely lightweight write access is relevant, if data provided by the high-priority audio context must be delivered to a low-priority thread. For example, this constellation appears in the task issuing routine transmitting simulation queries to the scheduler, which is extensively executed as tasks are generated by the up-sampling interpolation routines operating at audio frame rate (cf. Section 4.4.2). A concurrent queue internally decouples the pushing and consuming of arbitrary data objects while providing a thread-safe access to these functions.

4.6 Digital signal processing

The sampling of time-continuous signals into time-discrete representations is as old as the digital age itself [OL07]. In the audio community, digital storage, digital streams and the processing of time-discrete audio samples has become indispensable. The term *Digital Signal Processing*, short DSP, is representative for the entire variety of hardware units and software components, like plugins for digital work stations that apply the concept of time-discrete audio sample manipulation [Smi07]. Audio sampling rates are selected well above the range of human perception and sample values are quantized using the floating-point representation with a bit-depth of 32. This improves the Signal-To-Noise Ratio *(SNR)* for small values, as acoustic DSP units mostly operate between -1.0 and

1.0. At the cost of increased quantization noise for extremely large numbers, in-software clipping is avoided in case this range is exceeded [Smi+97].

In the context of auralisation, DSP is engaged to generate sound pressure signals for the physical generation of virtual acoustic environments (rendering) using electro-acoustic transducers such as loudspeakers and headphones (reproduction). The desired reproduction usually dictates the rendering format, since the stream properties must match. For example, a binaural stream is adequate for headphones and transaural Cross-Talk Cancellation *(CTC)* systems, while HOA or channel-based formats like VBAP are suitable for large loudspeaker arrays (see also Section 4.8).

A DSP network consists of various processing units, which can be divided into three classes: sources, sinks and input-output modules. Sources do not operate on an incoming audio stream but provide an output stream. For example, a file loader transfers audio buffers frame-by-frame to make the content of a pre-recorded audio file available in a block-based audio processing chain. Also, fundamental units to synthesise random signals (e.g. Gaussian white noise) or repetitive signals (e.g. sines, saw tooth pattern) represent audio sources that are helpful for artificial sound generation. Sinks within an audio DSP chain are employed, if the audio stream is not manipulated but analysed or stored to the hard drive. However, in practice, these modules can be implemented without restriction as an input-output module that simply forward the incoming stream to the output while observing and transforming the data flow for the given purpose (e.g., as a loudness probe). The most important DSP modules are manipulating the incoming audio stream and provide a modified output stream. To design a comprehensive network of DSP units that realises the auralisation of a dynamic urban environment providing the acoustics modality for a real-time VR application, a set of required units is briefly presented.

4.6.1 Fundamental operations

The fundamental operations to manipulate a digital audio stream are scalar multiplication and summation of floating-point data types. In the context of auralisation, this set of operations is heavily used, for example, to apply the spherical spreading amplitude gain and implement the superposition of different wavefront signals at the receiver location.

On a commercial PC, these operations are performed by the CPU in an almost negligible fraction of a second. Nonetheless, because these operations are excessively applied, it is reasonable to make oneself familiar with the mechanics and to economise on the usage as far as possible. To theoretically quantise and compare the calculation effort of algorithms, the routines are broken down to abstract operations by combining one multiplication and one addition and regard them as a unified, atomic procedure called *Multiplication-And-Addition* (MULADD). The total number of such MULADDs can be counted or roughly guessed. With knowledge on the complexity class of an algorithm and the scaling behaviour, an estimation of the processing effort with respect to the available resources can be drawn.

The estimation of resource consumption is essential for the design and implementation of an auralisation system. For offline rendering applications, the overall execution time can be optimised. For real-time auralisation systems that impose strict rules on the time budget assigned to the execution of the DSP network in the audio processor, optimisation of any kind is advisable in order to maximise the rendering capacity (e.g., maximum number of propagation path or sound sources).

4.6.2 Convolution and filtering

Performing a *convolution* on a digital audio input stream is a mechanism to modify the content of a signal according to a Linear Time-Invariant (LTI) system [OL07]. In acoustics, many processes are considered to fulfil conditions that the scaling of the input results in an equally scaled output and equal behaviour at different observation points in time is guaranteed. If sound propagation can be modelled based on these assumptions, it is possible to describe the transmission characteristics from a source to a receiver entirely by a function that incorporates the corresponding modifications. For example, the acoustics of a room can be acquired by the IR, a function of time, including the specific reflection and reverberation pattern as well as the source and receiver characteristics of the setup. Furthermore, the Fourier Transform *(FT)* is applicable and the system can be equally described by the TF, a function of frequency. Transformation between the two representations is performed by the Fourier integral. Because acoustic processes (in this context) are described by the real-valued, input/output pressure signals, the function remains real and the transform represents a complex-valued function that corresponds to the amplitude and phase modification of frequencies

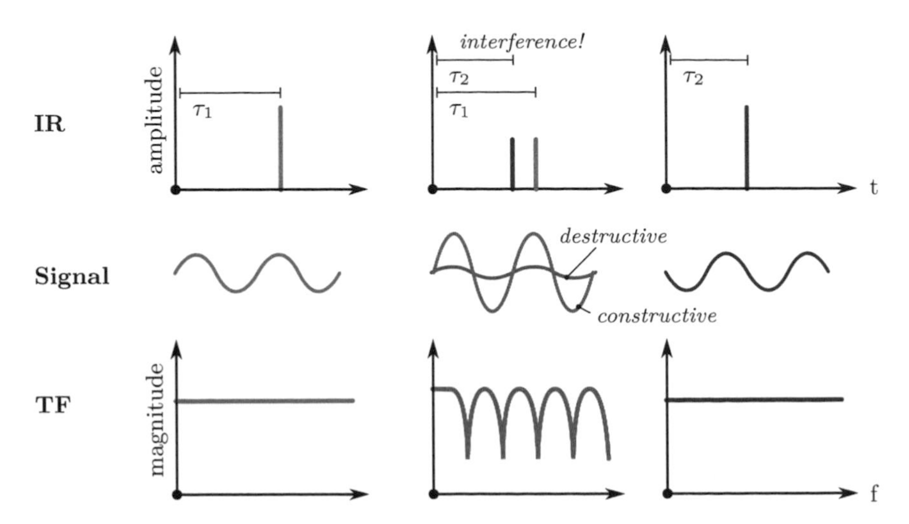

Figure 4.4: FIR filter exchange exhibits a comb filter due to an interference pattern, if two equal IRs represent a significant delay difference.

applied by the system [OL07]. Because the function is real-valued and causal, the transform is symmetric and it is sufficient to regard a single-half of the spectrum plus the Direct Current *(DC)* value.

The feature to apply a frequency-selective modification to an input signal makes convolution an essential processing unit for auralisation (i.e., FIR filtering). The property of full amplitude *and* phase control is often required, for example, to coherently superpose waves in room acoustic simulations, or to represent the inter-aural time and level differences of an HRTF for single frequencies. However, considering dynamic environments, the attribute of time-*in*variance forces to apply filter coefficient exchange over time. This approach inherits an intrinsic risk to cause audible artefacts, which is particularly prominent for motion adaption. Figure 4.4 visualises a filter exchange method that smoothly blends two convolved streams with relatively similar IRs, which are only shifted in time [TF07]. Unfortunately, using time-shifted impulses to represent delayed wavefronts in an IR is highly inefficient for single paths, because the empty initial range of the IR must be processed. On top of that problem, switching between two similar time-shifted IRs leads to a problematic comb-filter effect during filter exchange, which shifts collapsing frequency magnitudes into the lower frequency range with increasing relative distance – a circumstance that occurs regularly in urban environments

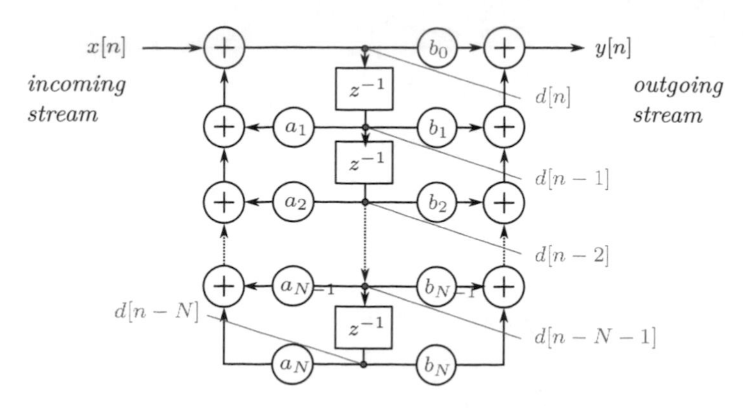

Figure 4.5: Flow diagram of a digital filter unit operating sample-by-sample in the time-domain.

with fast-moving sources and is clearly perceivable as artefact for signals with tonal components.

Apart from convolving IRs, *filtering* signals with recursive feedback loops achieves a similar result regarding the magnitude of the target spectrum, however, the phase control is dropped. Because many acoustic models, like absorption phenomena, only specify the magnitude spectrum, it is reasonable to employ IIR filtering, instead. IIRs describe purely auto-regressive processes that are highly efficient compared to FIR convolution, however, can be unstable and require a coefficient design method that approximates the desired magnitude spectrum. Figure 4.5 shows the flow diagram of a filter unit, which only consists of multiplication and delay operations. The direct implementation of an FIR convolution can be achieved by considering the blue tracks, while a pure IIR filtering is only considering the green tracks. While efficient ways regarding the MULADDs exist that implement time-domain FIR convolution by block-based algorithms that transform into the frequency domain for long IRs, IIR convolution is realised in the direct form, because only a few delay stages are required for a sufficient result (order 4 to 10 is proposed for propagation path attenuation filtering, cf. Section 5.4).

4.6.3 Delay lines

Delaying a signal is an important aspect in sound propagation modelling and auralisation. In contrast to light, sound waves travel at a relatively moderate speed through the medium, and the human auditory system is very well aware of slight time delays in a coherent acoustic signal [Bla97]. A *delay line* is a DSP unit that delays an incoming signal for a given time τ. The digital version is called tapped delay line, because every single sample is stored in a buffer or sequence of taps. A *fixed-length*, discrete delay line pushes incoming samples to the front and pulls output samples from the buffer end, which effectively delays the signal corresponding to the number of taps divided by the sampling frequency. This corresponds to a delay that is bound to integer multiples of the sampling rate,

$$L = \mathrm{round}(\tau) \cdot f_s$$

If a non-discrete time raster below the sampling rate is desired, the output of the delay line must be interpolated to produce a sub-sample resolution, called Fractional Delay (FD).

Fractional delay line

Fractional delay lines are able to delay a digital input signal without restriction of a discrete time raster. For a sub-sample resolution, the delay line must employ an interpolation on the time-discrete signal. An accurate implementation that recreates the time-continuous and value-continuous original (analogue) signal from the digital representation without degradation must apply an ideal low-pass filtering, as discussed in [Laa+96]. Therefore, a perfect recreation is only theoretically possible, as it requires the convolution of an infinite IR. Bearing real-time auralisation in mind, efficient non-linear routines to approximate the original signal are considered, instead. The side effect of non-linear signal interpolation is non-linear interpolation noise, which appears in the frequency domain as attenuated overtones due to missing higher-order polynomial terms. Depending on the order of the interpolation and the content of the signal, these disturbances are expected to be negligible for most applications.

To carry out an assessment of practical relevance on the expected signal degradation introduced by the interpolation routines, the performance of pure tone signal interpolation is evaluated. They easily reveal discrepancies of the employed algorithms as well as implementation errors during development. In contrast to sine signals, incoherent random noise signals represent the most robust signal type with respect to artefacts, as the excitation of the entire frequency range effectively invokes spectral masking [FZ07]. Relevant sound source signals that are practical for outdoor noise auralisation are anticipated to consist of both tonal and noise content. Hence, an informed decision must be made for each real-time application individually to meet the requirements between auralisation quality and quantity of sound sources (more specific, propagation paths) that can be integrated. Two cost-efficient interpolation routines have been implemented that appear suitable for outdoor sound auralisation purposes. Linear interpolation considers the nearest two data points and approximates the intra-sample value based on a linear function evaluated at a fraction, yielding

$$y_{interp}[n] = (1 - a) \cdot x[n] + a \cdot x[n + 1] \tag{4.1}$$

where $x[n]$ and $x[n + 1]$ are two adjacent sample values. This rudimentary approach has the advantage of low-latency, as the processing must only wait for one single sample to make the base value $x[n + 1]$ available. $a \in [0, 1)$ is the subsample fraction representing the residual part of the desired delay τ in seconds following

$$a = \tau \cdot f_s - \lfloor \tau \cdot f_s \rfloor \, . \tag{4.2}$$

Figure 4.6 depicts the re-sampling of a high-frequency tonal signal with a resampling ratio of $r = 2/3$ (i.e., producing 3 intermediate samples for 2 base samples) using a linear interpolation routine. Higher order routines are applied in a similar way and achieve significantly lower interpolation noise at the expense of an elevated computational effort. Also, they increase the delay, as they require more overlapping samples to cover the range of base samples for the desired interpolated sub-sample. A second class, the Spline interpolation, has been realised for a reasonably improved quality, if desired. Spline routines align well with the block processing nature of digital audio streams (cf. Section 4.7.1). Firstly, an intermediate result is calculated for an entire segment. Subsequently, the evaluation of interpolated samples uses only the intermediate result, which is generally fast. However, segment-wise Spline interpolation does not surpass linear interpolation in terms of a per-sample calculation effort regarding the required MULADDs.

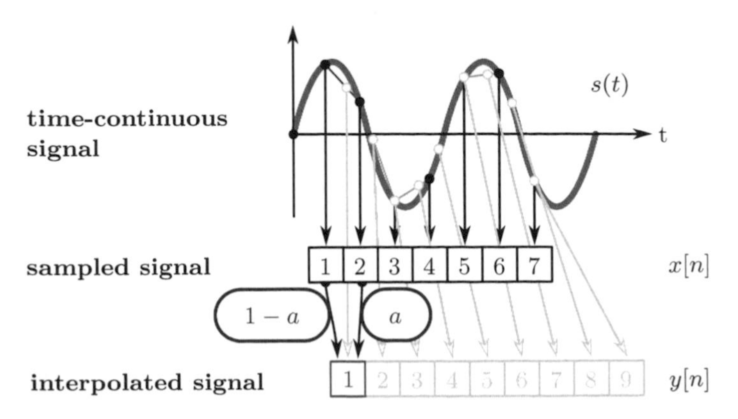

Figure 4.6: Linear interpolation of a high-frequency sine signal determining inter-sample values with a re-sampling ratio of $r = 3/2$ after Eq. 4.1.

Interpolating variable delay line

A digital signal processing unit that applies a delay and is capable of adapting to a dynamically changing delay parameter is called a Variable Delay Line (VDL) (or variable-length single-input single-output delay line). VDLs are inherently time-variant and perform an interpolation of the buffered audio stream, if the delay value is changed. Depending on the algorithm, the interpolation routine of a VDL requires a certain range of base samples (overlap areas). Figure 4.7 depicts the flow diagram of a VDL based on a digital buffer. New samples on the input are pushed to the front of the delay line and the residual samples are shifted. A variable read cursor (purple) picks up samples from the buffer corresponding to a time-dependent delay parameter, $\tau_{\text{delay}}(t)$. With the required overlaps, an interpolation is performed on a range of samples and the output is produced sample-by-sample. If the delay parameter is decreased, the variable cursor moves closer to the beginning of the buffer, and vice-versa if the delay is increased.[31]

[31] As long as the delay modification does not exceed the speed if incoming samples, the buffer ranges are always valid (i.e., velocities are lower than speed of sound). However, the buffer length grows with increasing delay values, which must be minded if realised by a ring buffer of limited capacity.

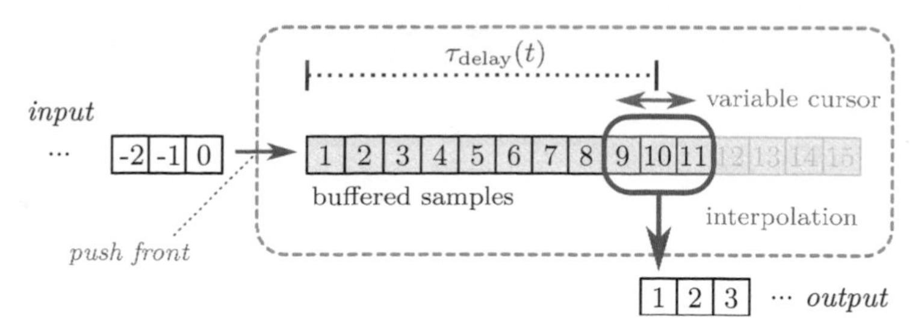

Figure 4.7: Single-input, variable-length delay line with a single, interpolating read cursor.

Single-input multiple-output VDL

An extension to the single-output VDL that is of particular interest for outdoor sound auralisation is the multiple-output variant, the Single-Input Multiple-Output Variable Delay Line *(SIMO-VDL)*. It features an arbitrary number of interpolating read cursors economising on the memory footprint, as the buffer is re-used to provide multiple output tracks. Apart from this aspect, it behaves exactly like the VDL including the possibility to apply an interpolation function to the output buffer, which is performed per read cursor individually. Considering the auralisation of dynamic outdoor environments, each moving sound source instantiates one SIMO-VDL, feeding the input with the source's signal, while picking up a delayed version using one read cursor per individual geometrical propagation path to model a) the propagation delay and b) the Doppler shift during read-out [SV18] (cf. Section 4.7.2). The SIMO-VDL represents a central unit of the proposed auralisation DSP network.

4.7 Audio rendering

Audio rendering is an essential part of an auralisation framework. It produces a digital output stream using DSP methods to realise an interactive, virtual acoustic environment. The output is intended to be processed and presented for one or many listeners by a subsequent reproduction unit, which commonly dictates the required format (cf. Section 4.8). Because the focus is laid on real-time outdoor auralisation for VR applications, an audio rendering approach is presented that builds on the deterministic pathfinder simulation method from Section 3.6 and produces a binaural output[32].

The task of an audio rendering module embraces the management of a substructure that represents the virtual environment and the parametrisation of DSP units that drive the rendering process. Additionally, in a real-time auralisation framework with a simulation scheduler and an auralisation timeline, propagation simulation results must be interpreted and partly interpolated in order to create the desired acoustic effect, such as the Doppler shift (cf. Section 4.3 and Section 4.4). The proposed realisation makes use of multi-threading and acts as a mediator between different contexts, namely the threads for scene handling, simulation scheduling and audio processing using methods described in Section 4.5.4.

4.7.1 Audio context

The heart of an interactive real-time auralisation system is the *audio context*, an application thread that produces the output stream and acoustically renders the virtual environment. This audio thread is driven by a high-prioritised audio call-back function, which is executed constantly at a user-configured rate and which is time-critical. In practice this means that the accumulated run-times of every operation within the function may never exceed the time budget, otherwise acoustic drop-out artefacts are inevitable. Depending on the audio architecture, the timing of the audio stack is highly precise, which is very important for a accurate parameter up-sampling.

The signature of the call-back function provides access to incoming audio data from the interface's inputs (e.g., a microphone, line-in, software loopback) and expects that the implementation fills an output buffer with audio data to be forwarded to the interface's outputs (e.g., D/A converters, amplifiers for headphones and loudspeakers). In practice, commercial audio interfaces are configurable regarding the desired sample rate (samples processed per second, units of Hertz) and provide a pre-defined list of buffer lengths that are processed within one call-back loop (samples per buffer[33], usually an integer size). For interactive processing, high-quality sound cards are available that provide block sizes as small as 16 sam-

[32] A re-design to match other output formats like VBAP and HOA is conceivable and requires a substitute spatialisation module.

[33] also referred to as block length, block size, (audio) buffer size, (audio) buffer length or (audio) frame

ples at sampling rates up to 192 kHz. Built-in sound cards usually offer a block size starting at 256, 512 or 1024 samples, whereas selecting higher block sizes is generally possible. Intermediate audio layers may provide features that combine blocks or perform automatic re-sampling to match client and hardware sampling rates. These audio mediation frameworks produce additional latency and draw processing power from the audio callback function, which is not desirable for a complex real-time audio application.[34]

The division of an audio stream into blocks of fixed length is required by the audio hardware and comes with the drawback of additional latency. A bigger block size forces a longer suspension of the interface to accrue the processed samples in order to forward them to the audio device. However, the time boundaries of the call-back function are refreshed at a lower temporal resolution, which increases the time budget for the processing units in the audio stream. The block time scales linearly with the block size and reciprocal with the sampling rate of the time-discrete system, yielding the block time formula

$$T_B = \frac{L}{f_S}$$

with L the number of samples processed per audio call-back execution and f_s the sampling frequency in Samples/s. For an interactive auralisation application, a practical value lies approximately at $T_B = 1.2\,\text{ms}$, which can be achieved using a block size of $L = 128$ samples at a sampling frequency of $f_S = 44.1\,\text{kHz}$[35]. According to Nyquist's second theorem, frequencies up to 22.1 kHz can be represented in the digital domain, which suffices the human auditory system [OL07; FZ07].

4.7.2 DSP network layout

One of the cornerstones of a successful real-time auralisation application is the layout of a network of DSP units that a) generates source signals, b) applies propagation effects and c) creates a spatial acoustic impression that reflects the virtual

[34] For example, Steinberg's® ASIO™ audio driver architecture is machine-oriented and is available for most commercial high-end audio interfaces, which makes it distinctly suitable for real-time auralisation.

[35] However, processing at the next common frequency value of 48 kHz is often found in audio hardware and software systems, giving reasonable leverage on latency at the expense of a tenable higher computational load.

acoustic environment. The requirements are manifold and context-dependent. A layout is presented that is relatively universal yet configurable with respect to a compromise between quality and quantity. It has the capacity to process a limited number of geometrical propagation paths between several dynamic sound sources and one or several dynamic sound receivers on a commodity PC in real-time. Each sound source generates a unique source signal and the content is transmitted via different paths to the receiver. The layout is tailored for binaural rendering[36] based on a combination of measured Head-Related Impulse Responses *(HRIRs)* using FIR convolution. A particular feature of the proposed DSP network is the intrinsic ability to adapt for dynamic (time-variant) scenes using sound generation and SIMO-VDLs, as dynamic, fast-moving sound source are expected in urban environments.

Sound source signal generation

In digital auralisation systems, source signals are represented by a time-discrete series of amplitude values that are directly proportional to the emitted sound pressure waves (DC free signal relatively to a static pressure value). The content of the signal is theoretically irrelevant for the processing chain. As an example of best practice, source signals should generally meet two conditions to avoid undesired side effects, and should ideally contain meta information to calibrate for absolute values[37]: Firstly, care must be taken that the respective amplitude of the provided signal does not exceed digital values of ± 1.0 in order to avoid clipping (which however is not of great concern if employed DSP units in the processing chain operate on floating-point values, cf. Section 4.6). Secondly, the content must contain a sufficiently high SNR [SSN04]. Stochastic background noise in the input signal can be dramatically amplified in the auralisation engine with very few possibilities for improvement.

Source signals are provided by the content designer of the virtual environment, for example, as a pre-synthesised loopable audio snippet or a recorded audio track. One advantage of loading audio tracks is the possibility to provide content in a straight-forward fashion. They are obtained from real-world recordings or sound clip databases. However, one major disadvantage of audio files is the lack

[36] The network can be remodelled to produce other formats, like HOA and VBAP, requiring an alternative implementation of the final stage, which accomplishes the directional cues for incident wavefronts at the receiver (spatialisation).

[37] In VA, per default, a signal with a *RMS*-value of 1.0 corresponds to an ideal point source that generates 94 dB re 20 µPa in 1 m distance.

of flexibility, which does not align well with outdoor source behaviour. Theoretically, pre-determined signals are only valid for static objects that do not change the emitted signal by external influences over time. Per definition, they are not suitable for sources that emit a dynamic signal depending on the internal state, a property that is of high relevance in virtual interactive environments.

Considering an urban scenario with many dynamically moving traffic vehicles and bearing the requirements for an input signal in mind, it appears more reasonable to synthesise source signals dynamically. For example, a car engine has many internal states and is subject to time-variant engine load, gear, acceleration and deceleration. Even a large database of recorded sounds and a sophisticated switching algorithm is not able to produce such a signal coherently. To overcome this problem, approaches that generate corresponding sounds in real-time are a promising method. For example, the concept of physics-based sound generation is built on methods of interactive sound design and presents a valid option that consequently uses artificially produced signals [Smi96; Far10]. In turn, the sound of a complex dynamic system is difficult to reproduce with a sufficiently high quality. This problem counteracts the level of authenticity and realism as it depends on the agreement and accuracy of the physical model. Another conceivable approach is based on analysis and synthesis of sounds with the help of recordings that provide a sufficient basis for interpolation [PBH16]. However, this method is limited to the laboriously established data and becomes infeasible if a variety of different vehicles is desired with numerous dynamic states.

The source-related sound power density spectrum requires special attention in the context of urban acoustics, because it provides a reliable importance indicator. Theoretically, the sound power spectral composition can be observed in real-time. In practice, it is reasonable to provide an envelope estimate of the sound power spectrum a-priori, as the designer of a virtual environment can easily categories the sources of the scene. This way, as an example, one can distinguish between a pedestrian who talks and produces footstep sounds that are only relevant in close vicinity and a heavy duty truck that is perceivable across an entire district. Considering acceleration methods for sound propagation simulations (cf. Section 3.6.8), pre-determined statistical values are valuable for an intelligent priority algorithm that classifies the relevance of a sound object with respect to other sound sources. Based on this data, the acoustic contribution to the perceived total sound field at a receiver location can be qualitatively predicted (e.g., by taking spectral masking into account). Consequently, irrelevant sound sources are excluded at an early stage from both the resource-heavy simulation

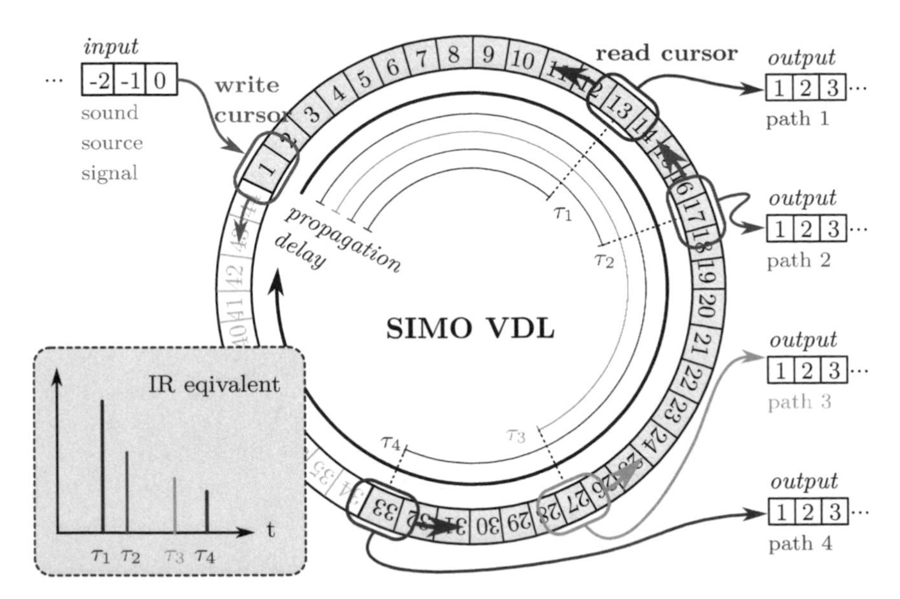

Figure 4.8: SIMO-VDL based on a ring-buffer implements propagation delay and Doppler shifts per source-receiver pair, exemplary for 4 different propagation paths at delays τ_i).

and the auralisation processing, which significantly improves the scalability of overall system (cf. Chapter 7).

Sound propagation processing

The rendering of dynamic, outdoor scenarios requires a time-variant DSP network. This requirement precludes a purely convolution-based approach that represents all individual geometrical propagation paths as one combined IR per source-receiver-pair for reasons discussed in Section 4.6.2 (cf. also Figure 4.4). Instead, a DSP approach is suggested reproducing sound propagation for individual paths that is similar to considerations mentioned in publications by Lokki et. al. and Wenzel et al. [LSS02; WMA00]. The schematic layout is depicted in Figure 4.8. Per sound source entity, a SIMO-VDL is used to realise the propagation of sound information in the medium. The input of the delay line is fed segment-wise by the sound source's pressure signal without modification at the system's block

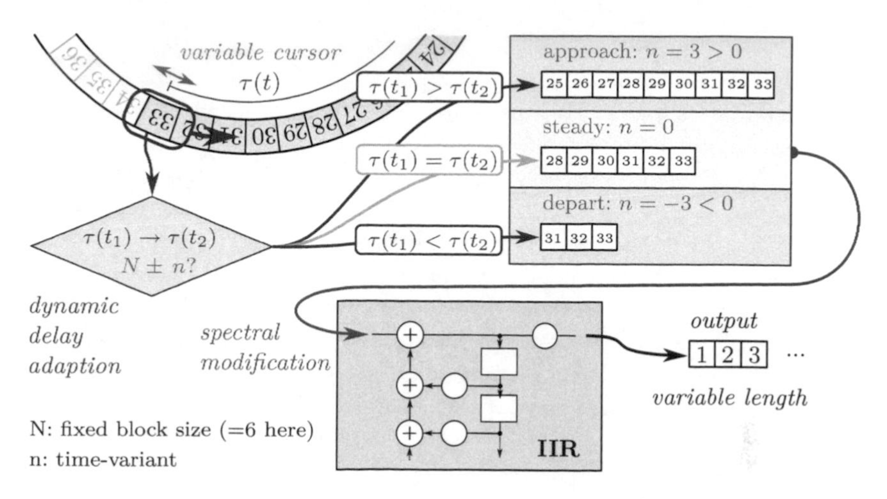

Figure 4.9: SIMO-VDL read-out and low-order IIR filtering of odd-sized output buffer for a fixed system processing bock size of $N = 6$ (cf. also Figure 4.8 and Figure 4.5).

length.[38] Per individual geometrical propagation path, one cursor is instantiated that performs a read operation at the corresponding delay value. During real-time auralisation, as long as the propagation path is valid, this delay parameter changes according to the up-sampled relative movement of source and receiver. Consequently, conveyed blocks of samples are made available as a sample buffer of variable length when adapting for dynamic delay changes, as depicted in Figure 4.9 (purple colours). If a geometrical path length is contracted, the delay value decreases and more samples must be read to move the cursor closer towards the write cursor (approach). In contrast, if a geometrical path length is extended, the delay value increases and less samples are available as the read cursor moves farther from the write cursor (depart). Performing a re-sampling of the odd-length sample buffer by interpolation to match the system's block size effectively realises a squeezing or stretching of the original source signal, as visualised in Figure 4.10 for both cases. Hence, a shift in frequency is perceivable according to Doppler's theorem. The interpolation, however, is integrated in the binaural spatialisation processor for efficiency reasons [SV18] (cf. Section 4.7.2).

With the help of the auralisation timeline, the source movement regarding the outgoing wave direction at the corresponding time $t - \tau_{\text{prop}}$ in the past and the

[38] The source signal is provided by loading samples from an audio file or by interactive signal generators, cf. Section 2.6.

current receiver movement regarding the incident wave direction is combined to a single Doppler shift value. This simplification is physically not correct, because the emitted signal with the source-induced Doppler effect is subject to frequency-dependent attenuation along the propagation paths, while the receiver-induced Doppler shift is effective on the observed signal in the medium after propagation. To solve this issue, a scaled attenuation spectrum must be considered that depends on the source-evoked frequency shift. Filtering is applied before re-sampling the variable-length output buffer from the VDL read-out, as proposed by Stienen and Vorländer [SV18]. The suggested, final arrangement of DSP units for a per-path processing pipeline is depicted in Figure 4.9. The read-out routine applies a frequency-independent delay that is physically related to the geometrical path length and the transmitting medium, in which information can only propagate at a finite speed (i.e., air sound speed for urban environments). Further acoustic propagation phenomena occurring along each individual geometrical path, however, are frequency-dependent. It is unavoidable to incorporate a filtering class to reflect these alterations, which is modifying the sound source's signal during transmission, namely attenuation by source directivity, medium absorption, reflections and diffractions [Vor11; KB20]. These effects are applied by a purely-recursive low-order[39] IIR filter unit approximating the energetic attenuation spectrum per individual propagation path using a *Burg* design schema to model the auto-regressive coefficients [Bur68; De +96] (see green tracks of Figure 4.9 and also Section 5.4). The filtering is performed immediately after the read-out routine and is realised in direct form after Figure 4.5, wherein all values for b_i with $i > 0$ are zero. It is taken advantage that this straight-forward time-domain implementation can be executed sample-by-sample and is not bound to a fixed block length.[40] For this reason, the IIR filter unit is engaged to process variable-sized input buffers and produce the same number of odd-sized output samples provided by the read cursor of the VDL.

Theoretically, a re-sampling of the variable-length buffer to the system block size yields the acoustic sound field component of the corresponding propagation path at the receiver location, if no directional sound field information is required (i.e., a monaural result is desired). For simplification reasons, Figure 4.10 shows this case for a smaller buffer causing a shift to lower frequencies (increasing delay / depart), and a larger buffer resulting in a shift to higher frequencies (decreasing delay / approach). Because down-sampling causes aliasing for spectral content

[39] Satisfactory spectrum approximations are achieved at orders as low as four, however, filtering with 10 delay units achieves sufficient agreement (cf. Section 5.4.3).

[40] Filter modules operating in the frequency domain usually require a fixed-length transform size and cannot handle variable-length input buffers. In addition, they are not efficient for low orders [Wef15].

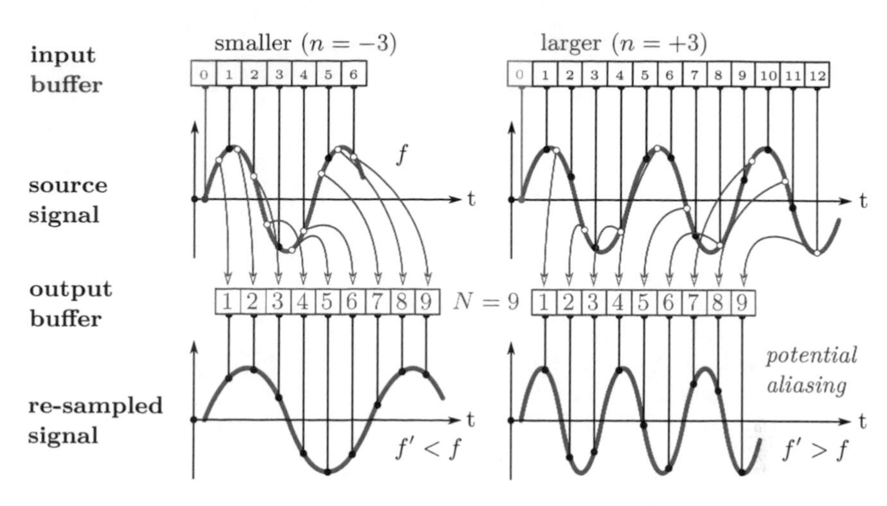

Figure 4.10: Doppler simulation by re-sampling odd-sized input buffers to the fixed system block length (cf. also Figure 4.9).

that is shifted beyond the Nyquist limit of the sampling rate, the preceding IIR unit is not only applying the scaled propagation attenuation spectrum, but also acts as a dynamic low-pass anti-aliasing filter, if required [OL07].

However, to produce a perception-related auralisation, a spatialisation is desired that considers the direction of incident wavefronts at the receiver location. An efficient binaural approach using HRIR datasets for many incident wavefronts based on a directional clustering is presented in the next section (cf. Section 4.7.2). The method applies a timing adjustment using FDs. To save calculations, all interpolation operations are centralised in the spatialisation unit regarding both re-sampling for Doppler shift and FD.

Binaural spatialisation

The directional processing unit generates a spatial sound format based on the incidence angle of all incoming wavefronts with respect to the receiver location. In case of a binaural observer and in contrast to locally stationary reproduction techniques using loudspeakers, the receiver orientation must be accounted for. Binaural streams maintain directional cues coded in the complex-valued HRTF, which contains frequency-selective amplitude and phase information for left and right ear in two separate time-synchronous channels. To congruently apply these

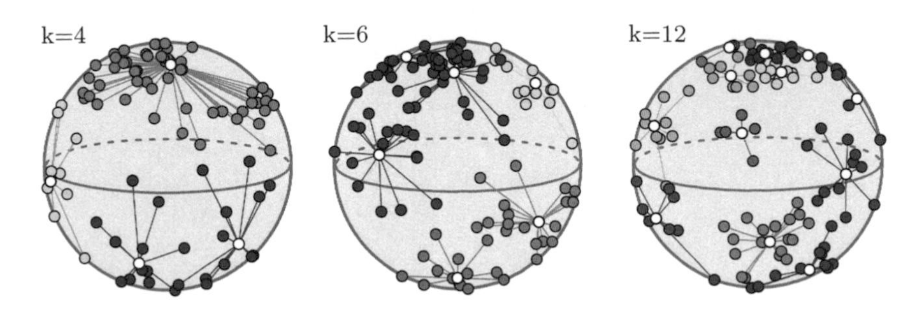

Figure 4.11: Assignment example of random incidence directions to clusters of k principle direction on a spherical surface.

properties, a two-channel FIR convolution of sufficient length is encouraged (e.g., processing an IR of 128 samples or more). In order to create a quality spatial impression, theoretically, each individual incoming wavefront must use one directional filtering unit, and the IR coefficients are obtained by querying an HRTF database of sufficient angular resolution (e.g., a measured, far-field HRIR with datasets at directions according to a Gauß grid of $1^\circ \times 1^\circ$ angular resolution).

As mentioned earlier, the employment of a costly two-channel FIR convolution unit per propagation path is considered undesirable as it rigorously limits the number of paths that can be rendered in real-time. To address this problem, a binaural directional clustering is proposed that consists of a fixed number of Principle Directions *(PDs)* that are clustering a subset of incoming wavefronts, as described by Aspöck et al. [Asp+19]. Each PD instance runs a complete and adequately long HRTF filtering using a two-channel FIR convolution. Figure 4.11 shows an example, where incident wavefronts are grouped (colored dots) in the vicinity of PDs (white dots) for 3 cluster orders. Signal streams from these paths are routed through the corresponding PD convolution unit, as they are expected to inherit similar directional features. The suggested approach possesses the appreciable advantage that a constant DSP load is imposed for the directional convolution effort, which is basically independent of the propagation path number. However, it comes at the expense that incident wavefronts are perceptually warped towards the principle directions resulting in a mismatch between actual and processed binaural direction. Additionally, the clustering algorithm introduces an overhead whenever a scene state change is issued,[41] because a re-evaluation of the directional grouping and the principle directions must be performed. The procedure to determine the PDs is based on an attempt to minimise the maximal

[41] More precisely, when a simulation result is made available by the scheduler, cf. Section 4.4.2

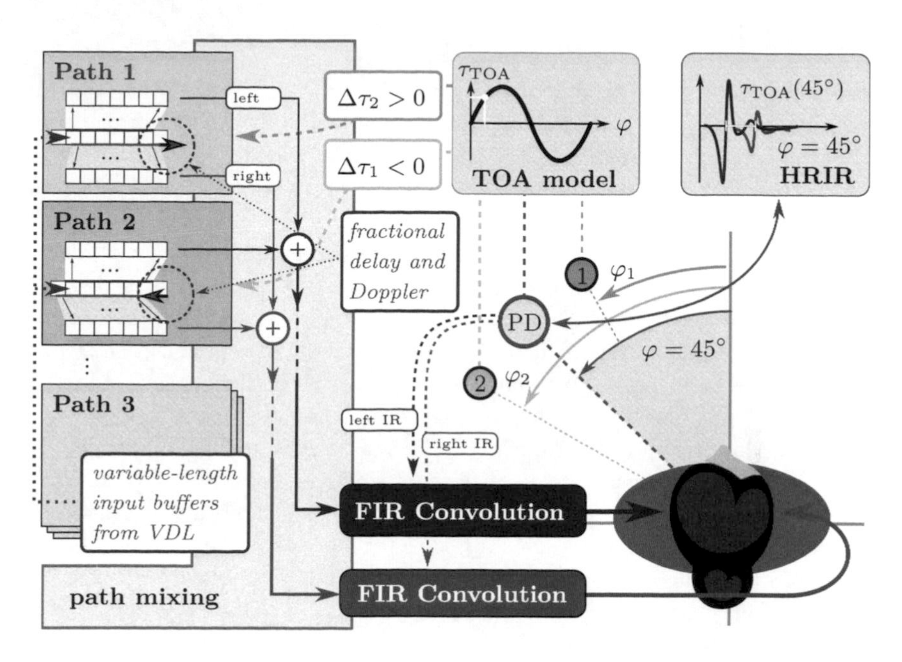

Figure 4.12: Spatialisation diagram using one HRIR convolution for the PD and mixing assigned directions while correcting for the ITD based on a Time of Arrival *(TOA)* model *and* applying the Doppler shift in a single interpolation routine per path and per binaural channel (left and right).

angular error of directions (feature space of clustering). Since this minimisation approach is an NP-hard traveling salesman problem, the calculation time is crucial and heuristic methods are reasonable [Vat09]. Therefore, an approach is realised that initially uses the farthest-first traversal in order to find k PDs with largest distance in the feature space [RSL77; TGD04; Moe+07]. Such a method promotes outliers, since, for example, a single path from the rear must definitely receive a cluster direction, especially in case of many frontal paths. Then, a k-means weighting scheme is performed that assigns closed-by directions and approximately minimises the relative angular error by shifting the cluster centre, which can potentially be extended to other scalar weighting metrics based on perceptual models. Measures are considered to compensate for sequences of clustering routine results upon scene updates that are significantly different. The realisation avoids irregular snappiness while maintaining a stickiness in order to provide a certain persistence in the distribution.

Figure 4.12 depicts the DSP layout of the binaural spatial rendering unit that realises the path mixing and concludes the DSP network layout of the proposed real-time rendering method designated for dynamic outdoor sound propagation. The final angular deviations in the horizontal plane of the listener are estimated by a TOA model to adjust the corresponding ITD error per incoming wavefront with respect to the processed HRTF from the PD in the DSP pipeline. To increase efficiency, the ITD correction feature[42] and the Doppler simulation is individually applied per propagation path exploiting that the interpolation stage of the VDL output has been postponed. For the presented spatialisation approach, a simplifying spherical estimation method after Ziegelwanger et al. has been chosen [ZM14]. However, it is conceivable to employ more sophisticated individualised models or derive a correction function by an ITD curve fitting algorithm that evaluates the HRTF database [Møl+95; HP06; NAS08]. According to the ITD model, the delay mismatch between the actual incident angle and the principle direction is reverted in the horizontal plane by performing *two* interpolating read-outs from the (filtered) VDL buffer, one for the left and one for the right ear. At the expense of a doubled interpolation cost, the approximate frequency-independent inter-aural propagation delay difference is taken into account for each path by integrating the value into the interpolation routine. Consequently, the number of required principle directions is largely reduced and valuable processing time is saved. The directional processing unit is configurable in terms of number of PDs and refining iteration cycles. Further, the implementation provides a fast receiver rotation update method that skips a re-calculation of the clustering state. If the user's head turns, it is sufficient to re-determine the HRTF dataset angles, adapt the ITD corrections linked to the steady PDs and perform a filter exchange. The algorithm is not able to compensate spectral properties of angle mismatches in the vertical plane (elevation).

4.8 Audio reproduction

The *audio reproduction* component represents the final stage of an auralisation framework. It embraces the processing and routing of the rendering output, a multi-channel audio stream, to the electro-acoustic transducers, for example, headphones or loudspeakers. Therefore, it is aware of the hardware setup and is commissioned to transform and transmit the incoming digital audio stream to one

[42] The ITD is understood as the delay imposed by the geometrical detour that a plane wave encounters in a medium of finite propagation speed, i.e., the TOA, which is frequency-independent in non-dispersive media such as air. It corresponds to the group delay of an HRTF.

or many users. The digital input stream therefore must be coded in a mutually matching format. To the reproduction unit, it is irrelevant if the incoming stream is a recorded audio file or a rendering output. A variety of different methods and configurations are conceivable, however, the decisive aspect is the desired physical playback constellation. Therefore, one differentiates between

- headphone-based reproduction and
- loudspeaker-based reproduction

with the essential distinction that a pair of headphones must be *worn* by the user, while loudspeakers are mounted at fixed positions. Considering an interactive real-time application, the reproduction system is either robust against user movements (e.g., VBAP, HOA) or realises adaption using motion tracking (CTC).

The employment of headphones is favourable for low-budget virtual reality systems, if the drawback of a potentially *intrusive wearable* can be accepted. Firstly, headphones assure a cross-over free transmission of the binaural audio stream to the left and right ear. Secondly, user motion does not affected the transmission making dynamic adaption in the reproduction stage obsolete. Further, headphones are relatively robust against outside influences and do not require a special treatment of the surroundings (e.g., considering room acoustics, equipment noise, insufficient insulation, etc.). And last but not least, the quality and effortless reproducibility of a headphone-based reproduction makes it an excellent choice for scientific user studies.

In contrast, loudspeaker-based reproduction systems are hardware-intensive and require an acoustically treated environment to minimise distorting reflections off walls. On the contrary, loudspeaker systems can be used to create an acoustic display for more than a single listener and are non-intrusive, which is advantageous in VR applications.

The presented auralisation system can only be accompanied with headphones or a dynamic CTC reproduction system, since it is designed for binaural rendering in the initial implementation. A binaural auralisation framework is predestined for applications using Head-Mounted Displays *(HMDs)* with headphones as well as *CAVE*-like systems using dynamic CTC via loudspeakers, as realised in the *aixCAVE*, the interactive VR system of the RWTH Aachen University [Cru+92; Len+07; Sch+10].

4.9 Summary

A real-time audio processing framework has been realised that is able to acoustically render virtual outdoor environments with dynamic, fast-moving sources and interactive receivers in a static built environment. The basic structure follows the initial concept of Section 3.1, schematically depicted in Figure 3.1, while more details are presented concerning the integration of a Geometrical Acoustics (GA) simulation scheduler, the management and interpolation of simulation results and the implementation of a DSP network driven by the audio process.

To comply with aspects of real-time auralisation constraints, it is proposed to employ different contexts with separable tasks regarding dynamic scene handling, acoustic simulation and audio processing. The approach introduces an auralisation timeline to record and up-sample scene updates and sort simulation results from a scalable scheduler instance in a chronological layered history. Based on the history data, a comprehensive treatment of propagation delays with the possibility to minimise latencies by approximate interpolating methods is achieved. A sophisticated Digital Signal Processing (DSP) network is presented that processes propagation effects such as spectral attenuation – including diffraction – and the time-variant Doppler shift for many *individual* geometrical sound paths in real time using low-order Infinite Impulse Response (IIR) filter units. Key feature of the suggested design is the implicit capacity to adapt to fast motion by introducing Single-Input Multiple-Output Variable Delay Lines (SIMO-VDLs) that congruently model propagation delays over. Since dynamic environments are considered, the presented layout consistently replaces the classical steady-state convolution approach by an appropriate time-variant alternative. Finally, a binaural spatialisation procedure is introduced that is based on a *directional clustering* routine at the receiver end and operates at a quasi-constant computational load. The method uses a fixed number of Head-Related Impulse Response (HRIR) convolutions and adjusts the Inter-aural Time Difference (ITD) mismatch using Fractional Delays (FDs) or each individual propagation path, while maintaining efficiency by merging Doppler interpolation and ITD correction into a single routine. The complete layout of the DSP network that represents the core of the audio rendering module is visualised schematically in Figure 4.13.

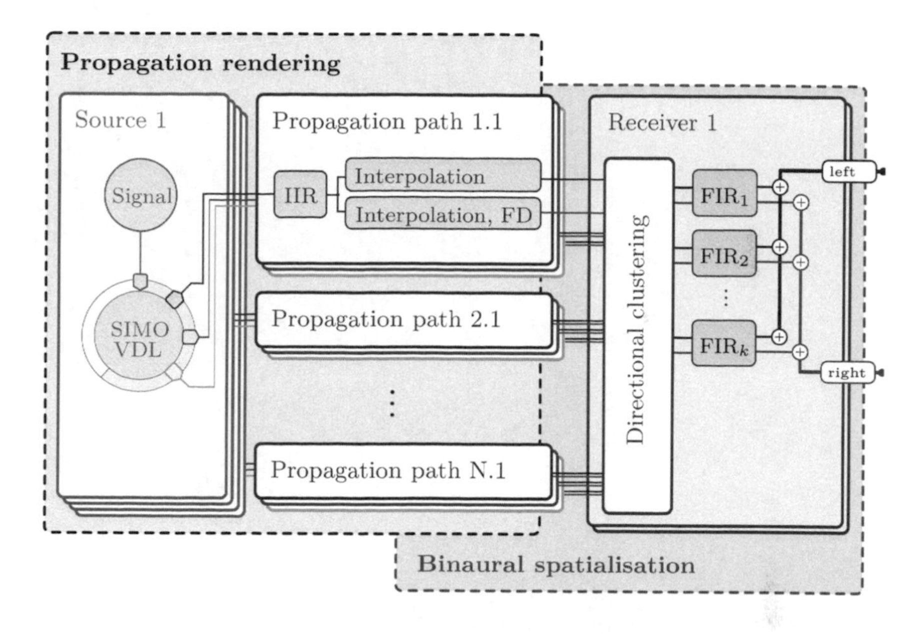

Figure 4.13: Summary of the proposed DSP network tailored for the real-time auralisation of dynamic outdoor environments. Each source N provides a signal that feeds a SIMO-VDL. Read cursors per source-receiver pair transfer a delayed propagation path signal through an IIR filter realising spectral propagation attenuation. After separation into left and right channels, interpolation applies Doppler shifts and FD for one channel to realise ITD correction in the binaural spatialisation. A directional clustering mixes paths and routes the accumulated signals through the k PD units, each running a full HRIR convolution. Finally, per M receiver, the binaural output is mixed.

5 Application and verification

5.1 Introduction

The presented real-time auralisation framework is intended to provide acoustics for dynamic outdoor VR environments. Depending on the desired goal and the available computational resources, the framework can be engaged for different purposes under real-time constraints. At the current stage, the scope of the realisation considers limited, relatively modest scenarios with a few sound sources, a built environment that is by design reduced to a manageable size and an output for one or two receivers that are rendered in parallel. Hence, feasible applications lie in the range of informal demonstrations for noise assessment and experimental user studies. Conceivable scenarios that can be executed on commodity PCs include street crossings, squares and residential districts containing buildings as rudimentary geometrical shapes. Distributed computing is proposed for the workers of a simulation scheduler to avoid interference with the machine that is also entrusted with the real-time rendering. This way, either the update rate or the resolution of the simulation scheduler can be increased, for example, by incrementing the reflection and diffraction order. Also, regarding complex scenarios, the potential bottleneck of the auralisation system is the capacity of the DSP network, as it can only process a limited number of propagation paths at low-latency, real-time rates. If concurrent access to computational resources is avoided, either more paths can be processed in the rendering stage or the resolution can be improved, for example, by increasing the IIR filter order and by enabling spline interpolation instead of linear interpolation for the Doppler shift and the FD in the spatialisation unit.

The appropriate presentation of an auralisation result is unquestionably an audio-video demonstration and the adequate evaluation of a real-time rendering is to experience the framework in a VR application. However, to provide evidence that the described system is feasible, an independent verification of the components

- propagation simulation using the GA pathfinder algorithm,

- acoustic modelling using the UTD method and
- auralisation by employing the proposed DSP network layout

is conducted. The performance of the components are validated on a commercial desktop PC without parallelisation to provide a robust result that allows the estimation of requirements for a desired purpose employing comparable hardware systems.

It is outside the scope of this thesis to deliver a verification of the manifold perceptual consequences of setup and configuration of the presented auralisation framework, as these specifics depend largely on the desired goal and the available computational resources. The results presented in this chapter are intended to support the decision process concerning the design of an experiment or demonstration in order to estimate the scale of the required hardware and the feasible auralisation resolution. In order to grasp the applicability for specific real-world problems using the proposed framework, preceding listening experiments must verify the individual configuration. Aspects like DSP quality regarding interpolation and filtering order, scheduling parametrisation, simulation abortion criteria, propagation path importance sampling and binaural clustering resolution must be carefully evaluated to find a suitable solution that delivers a plausible acoustic environment. In particular, to avoid negative side effects, it must be verified that the real-time rendering process does not cause artefacts due to an overload of the CPU or violation of the audio frame time budged. Additionally, it is important to guarantee that the simulation scheduler delivers a sufficient update rate for the given context.

To conclude the verification of the real-time auralisation framework for outdoor sound propagation, an example scenario is described that runs in an interactive VR environment on a standard desktop PC using an HMD with headphones driven by a graphics cards and an audio interface of commercial, budget quality. For the purpose of a technical demonstration, the choice of setup and configuration of the real-time auralisation framework is explained.

5.2 Propagation simulation

To investigate the performance of the geometrical propagation pathfinder algorithm described in Section 3.6, a street canyon scenario is considered and simulation results are evaluated from a source location to a receiver location.

Figure 5.1: Street canyon scenario with a source (red cube) and a receiver (blue cube). Between the entity locations, the direct sound path is not audible.

5.2.1 Scenario

The geometrical model uses 2 polygons and 70 edges. It represents a T-junction street canyon surrounded by buildings and noise barriers of different shape and height. A second building row following after the central block with a pitched roof is included to investigate the influence of backwards-directed sound behind a large occluder. Figure 5.1 depicts the urban setting and indicates the source and receiver location as a red and blue cube, respectively. All faces are modeled as acoustically hard.

5.2.2 Simulation results

Initially, a high-resolution propagation simulation is performed without any accelerations. An abortion criterion has been selected to keep runtime and memory consumption maintainable, which limits the maximum number of reflections and diffractions for a geometrical path to an order of five. After excluding acoustically problematic paths that violate the far-field condition, for example, if two interaction points are too close in distance, a total number of 1314 paths are determined.

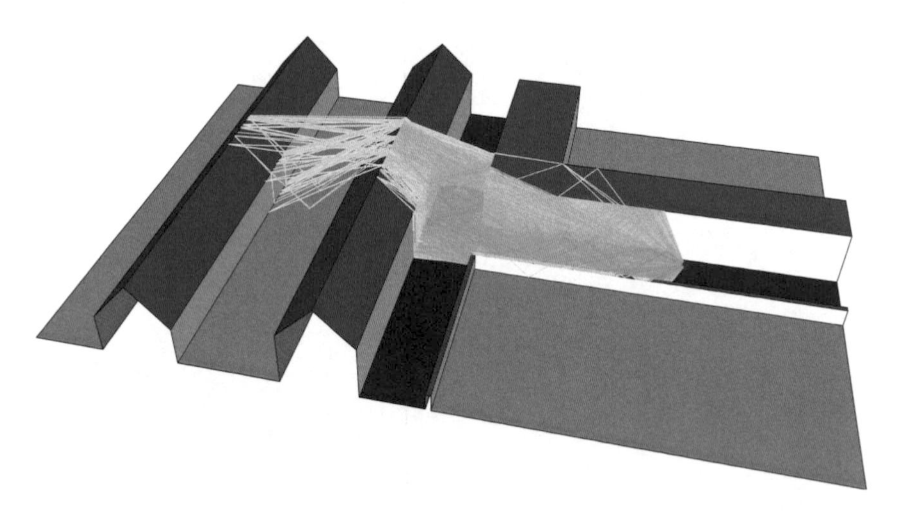

Figure 5.2: Street canyon scenario with all propagation paths up to order five. The individual line colour ranges from green to yellow, indicating the relative number of reflections and diffractions, respectively.

Figure 5.2 visualises the entire set of geometrical propagation paths for the described scenario. It becomes obvious, that the original number of possible source-receiver connections via the model geometry is overwhelmingly large and many contributions, for example, regarding the multiple-diffracted paths that are back-scattering from the second-row building, are unlikely to have a substantial impact on the perceived sound field at the receiver location. To evaluate the influence of order, the propagation paths are grouped by occurrence of reflections and diffractions. Figure 5.3 shows two three-dimensional stem plots with a mesh overlay above a horizontal grid of reflection and diffraction orders. As depicted, the propagation result neither contains a free path nor a first order reflection path, because the pairs $(x = 0, y = 0)$ and $(x = 1, y = 0)$ do not have a z value. Also, paths are capped at order five made evident by the diagonal edge of the mesh that spans from pure reflections $(x = 5, y = 0)$ to pure diffractions $(x = 0, y = 5)$. Considering the amount of paths per grouped order, only a small number of low-order pure reflections and diffractions is found. However, with increasing order, the exponential grows of the deterministic propagation algorithms is observed. Especially the number of paths with a few diffractions rapidly approaches high numbers, as indicated by orange-to-yellow mesh edges in the plot. The value rises approximately *linear* for z-values on the base-10 logarithmic axis scale for the path count. In contrast, considering the average path length per group in the

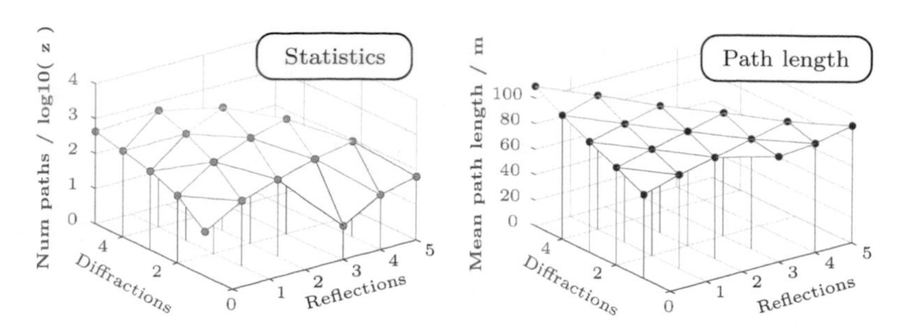

Figure 5.3: Propagation path statistics and average path lengths over orders of reflection and diffraction for all determined paths in the street canyon scenario.

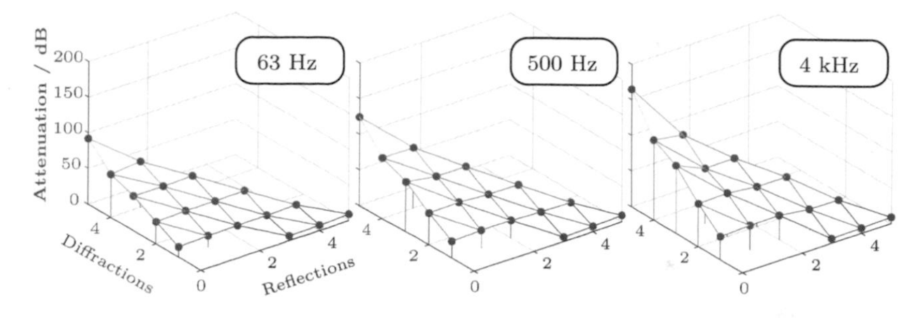

Figure 5.4: Average propagation attenuation of three representative frequencies over orders of reflection and diffraction for all paths in the street canyon scenario.

right plot, propagation distances are moderately growing with each reflection or diffraction order. Hence, a substantial acoustic impact from one order to the next higher order regarding the expected spreading loss is not apparent, as next-order paths are of comparable distance. Instead, for an improvement of the a-priori importance sampling, a diffraction model must be considered to account for the frequency-dependent attenuation of diffracted sound. In Figure 5.4, three representative frequencies at 62 Hz, 500 Hz and 4 kHz are selected and the propagation attenuation is determined based on the UTD model for each path individually. Again, an order grouping is performed and the contribution of each path is energetically averaged and referenced to the virtual free-path sound pressure level. Based on the averaged attenuation in dB along the z-axis, it can be seen that the dimension of diffraction drastically reduces transmitted sound. In contrast,

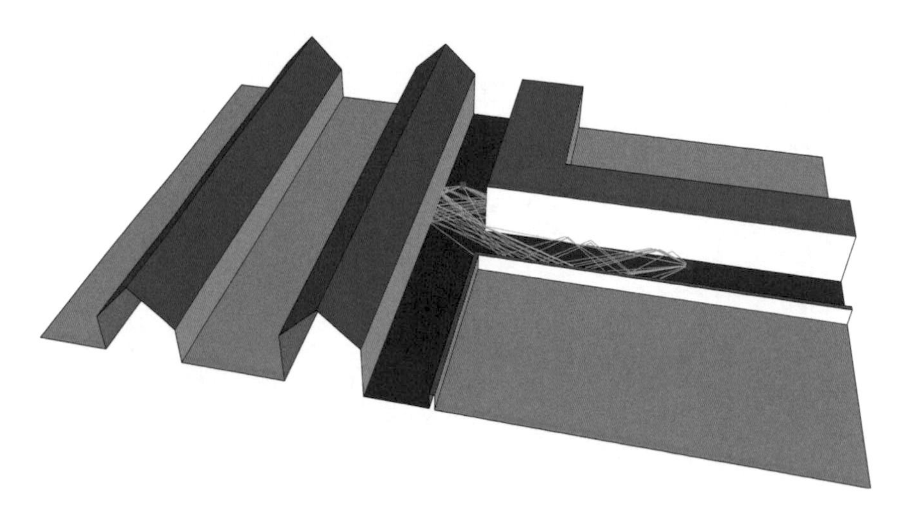

Figure 5.5: Street canyon scenario with the 20 propagation paths ranked energetically highest.

repeatedly reflected sound decreases insignificantly if the reflecting surfaces are modeled acoustically hard. The tendency of increasing attenuation with higher diffraction orders is observed for all representative frequencies. However, regarding the low-frequency range, these paths can only be considered irrelevant at higher orders for realistic urban environments, while high-frequencies are already largely attenuated at low diffraction orders.

In the context of real-time auralisation for outdoor situations and in the prospect of a limited amount of processable paths in the rendering stage, the question arises, which paths are acoustically most important and must be included. While multiple diffractions are apparently insignificant, highly reflected paths appear to continuously contribute to the overall sound field due to the largely non-absorbing surfaces. Bearing the sound field continuity requirement in mind, diffracted paths are equally important to compensate for appearing and disappearing direct and reflected paths in a dynamic situation. However, based on the results of Figure 5.4, energetically important diffracted paths that establish amplitudes as high as reflected paths only appear to consist of one or two orders. Hence, a general solution to the importance sampling tends to a constellation that integrates high-order reflections to account for the transmitted broad-band sound with a few, low-order diffractions that establish continuity and deliver diffracted sound into shadow zones that are otherwise unaccounted for.

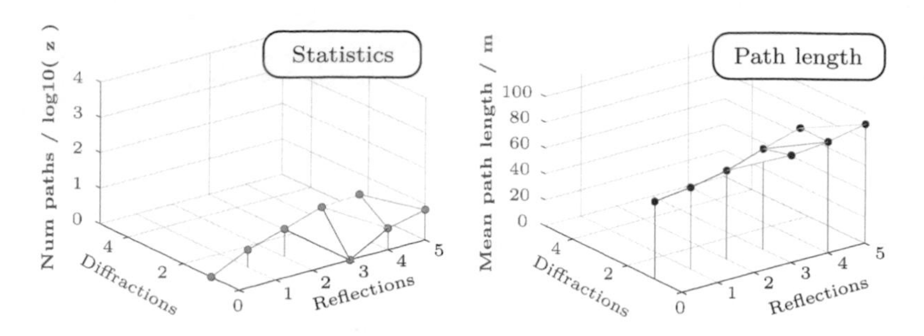

Figure 5.6: Propagation path statistics and average path lengths over orders of reflection and diffraction for the top 50 paths ranked energetically highest in the street canyon scenario (cf. Figure 5.3). Most paths consist of many reflections and a up to one diffraction component.

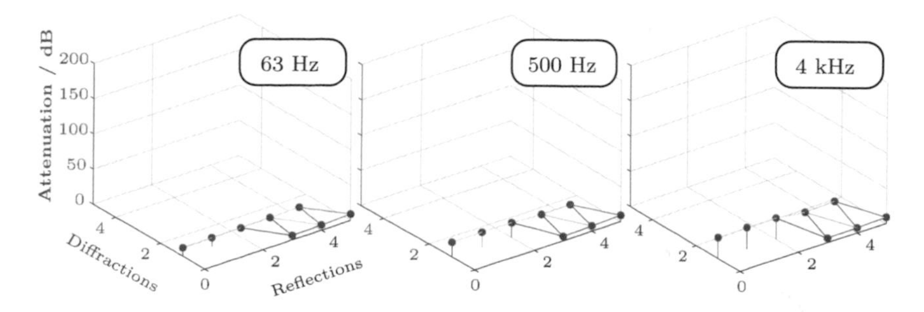

Figure 5.7: Average propagation attenuation of three representative frequencies over orders of reflection and diffraction for the 50 paths ranked energetically highest in the street canyon scenario (cf. Figure 5.4).

Figure 5.5 shows the 20 most important paths based on an energy ranking for a frequency range of 50 Hz to 20 kHz. The majority of paths include one or several reflections as they are not attenuated by absorption at the walls and do not take a large detour via the surfaces in this scenario. However, seven paths also possess a single diffraction including the purely reflected path of shortest distance from source to receiver via the edge of the vertical corner of the L-shaped building. This path is of particular interest, as it is required to compensate for the instantaneous appearance of the direct sound in case the source is moved across the junction and out of the shadow zone of this corner.

To estimate the development of importance sampling on a larger dataset, in the following, the 50 most important paths are considered to increase the validity when comparing statistical values.[43] Analogue to the full dataset, Figure 5.6 shows the number of propagation paths and the average propagation distance. The general distribution of the prioritised paths over the order grid demonstrates, that an asymmetry regarding the reflection and diffraction order is evident. A clear shift towards reflected paths can be seen, while diffracted paths of higher orders are entirely dismissed. Considering the average attenuation per combined order depicted in Figure 5.7, all reflected paths are included, while only a few paths with a single diffraction are found in the subset of the 50 top-ranked contributions. Therefore, for this scenario and the given locations of source and receiver, a propagation simulation determining paths of highest possible reflection order and a single diffraction order suffices to deliver the acoustically most important propagation paths in terms of transmitted broad-band energy using the UTD model.

5.3 Propagation model verification

Methods based on the GA principle largely neglect wave effects and therefore have to proof their applicability if employed for sound propagation modelling. Since the described real-time auralisation system is based on diffraction models, the prediction accuracy will be investigated in the following.

The pathfinder algorithm from Section 3.6 is able to determine propagation paths following the wavefront normal, while considering acoustic effects that refer to both specular reflections and edge diffractions. To conduct a comparison with measurements, a propagation simulation is performed and the resulting geometrical paths are transformed into the IR and TF by means of acoustic modelling after Section 3.4. In the literature, typical canonical situations regarding screening effects and wedge-shaped edge diffraction problems can be found, that predominantly target the acoustic model quality, for example, those by Schröder et al. [SSV11]. A more elaborate benchmark that considers a combination of reflection and diffraction on the one hand, and provides supplementary data to enable a high-resolution coherent reproduction is presented by Aspöck et al. as the Benchmark for Room Acoustic Simulations *(BRAS)* and will be used to verify the acoustic modelling of geometrical propagation paths [Asp+20; Bri+21].

[43] This particular value was chosen as it demonstrates most comprehensibly the correlation of diffraction order. Paths of next-higher diffraction order are included in the 150 most important paths, the third order already requires 400 considered paths.

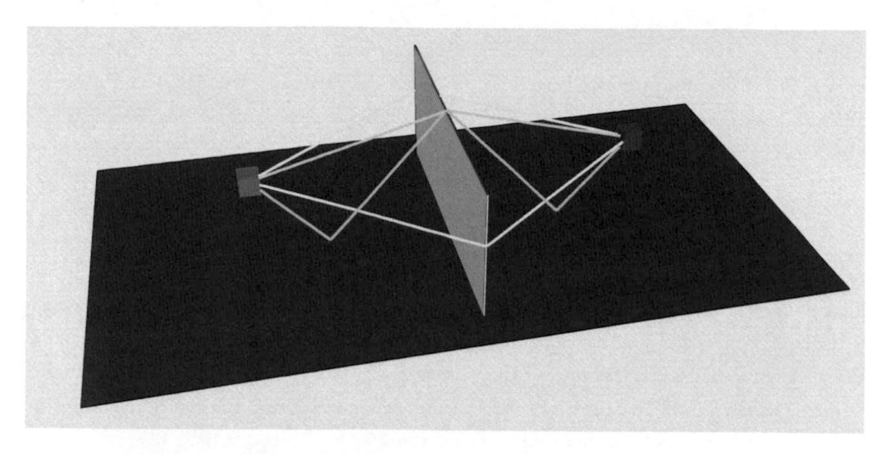

Figure 5.8: Visualisation of BRAS Reference Scene 5 (LS01, MP01) [Asp+20] indicating a selection of potentially relevant geometrical propagation paths (no transmission).

5.3.1 Benchmark for Room Acoustic Simulations (BRAS)

The BRAS is an exemplary project practising the principle of open science and offers sufficiently complete data to reproduce a few scenarios by simulation and compare the results with measurements [Asp+20]. Of particular interest for the evaluation of acoustic diffraction modelling is Reference Scene 5, as it consist of a large screen and includes situations of source and receiver locations that block the direct sound path. The provided scene description is given in thorough detail and the delivered acoustic measurement data is of high quality, allowing, for instance, the evaluation of the raw recorded IR that is of sufficient length and sampling resolution. It can be transformed into the complex-valued TF covering the entire audible frequency range. In addition, the supplementary data includes the directivity spectrum of the employed loudspeakers at a high spatial sampling resolution. The measurements were conducted in a hemi-anechoic room with acoustically treated walls and ceiling. The wedge-type absorbers greatly suppress reflections and the room dimension is sufficient to approximate free-field conditions for the upper hemisphere down to a frequency range of 100 Hz. Because a commercial loudspeaker has been used, in contrast to an approximately omnidirectional source, the corresponding directivity database is essential to reproduce the measured IRs by means of simulation with the necessary degree of accuracy, enabling to draw a conclusive comparison [ESV21].

Figure 5.8 qualitatively shows a visualisation of Scenario 5 with the loudspeaker location *LS01* and microphone position *MP01*, as described in the literature, by a red and a blue cube, respectively. It consists of a reflecting floor with a thin screen barrier in the centre blocking the direct line-of-side. Additionally, the figure draws the geometrical propagation paths determined by the pathfinder algorithm of Section 3.6. Paths with one diffraction and one or two reflections either in front of the screen where the source is located or behind the screen are included as green lines, while purely reflected paths are depicted in yellow. The screen appears infinitesimally thin, however, has a thickness of 25 mm. Consequently, in the pathfinder algorithms, the barrier is treated as a solid rectangular object and hence, to determine the shown wavefronts geometrically, at least a diffraction order of two is required. In the acoustic modelling, the wall is considered as an infinitesimally thin screen and the paths around the object are considered as a single diffraction. However, apart from an increased calculation effort regarding the order abortion criterion, the feature of the pathfinder algorithm to detect subsequent diffractions is mandatory if the original model is used and the screening plate is not manually collapsed to a single polygon. In fact, the acoustic modelling based on the geometrical propagation paths is executed by a MATLAB® script that considers distances between diffraction interaction points and automatically merges geometries, which is effective in this particular constellation (cf. Listing 5.1).

The resulting propagation paths exhibit a certain set of emission angles regarding the loudspeaker and immission angles regarding the receiver. While the employed microphone can be considered approximately omnidirectional and provides a frequency-independent sound pressure recording, the commercial loudspeaker shows an on-axis frequency response and considerable radial deviations. Figure 5.9 depicts the corresponding selection of directional frequency responses of the loudspeaker that are found in the propagation paths. It becomes obvious, that the directivity component requires attention in the acoustic modelling given the variance of several dB in general and some distinct irregularities in particular, for example, the destructive interference occurring at 3 kHz for the elevation angle of $\theta = -47.7\,^\circ$ (emission angle towards the ground reflection, cf. Fig. 5.8).

Another aspect that requires special treatment is the transmitted acoustic energy that propagates through the solid fibreboard. This contribution is included in the measurement, however, is not discovered by the pathfinder algorithm, as it is inherently unconcerned with sound inside solid structures. To achieve a practical basis for comparison, transmitted energy is *manually* added to the

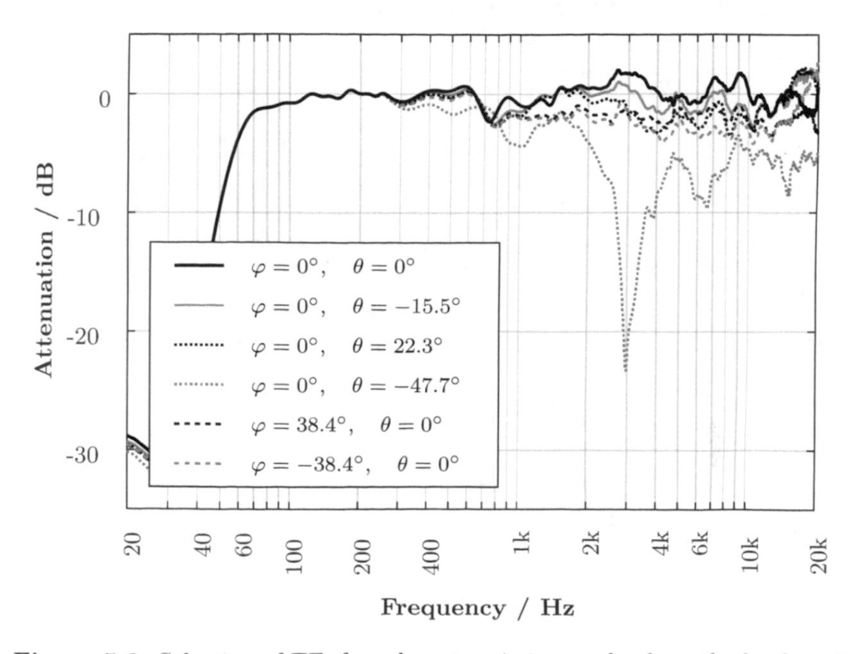

Figure 5.9: Selection of TFs for relevant emission angles from the loudspeaker directivity of BRAS Reference Scene 5, see [Asp+20].

simulation result of the scenario using the mass law formula according to Erraji et al. [ESV21].

To perform the comparison, the pathfinder algorithm is executed using the polygonal model of the scenario and the locations described in the BRAS documentation. The result is exported as a plain-text JavaScript Object Notation *(JSON)* file with a proprietary descriptive format that contains a list of all individual propagation paths. The file is then loaded by the MATLAB® class *itaGeoPropagation*[44]. For each path, the TF is calculated by applying the acoustic propagation models according to Section 3.4. An exemplary code snipped is shown in Listing 5.1, where the actual acoustic modelling is executed in the *run()* function, line 18, and applies the UTD model, but also supports, for example, the Maekawa method.

The result of the described procedure delivers the TF depicted in Figure 5.10 and the transformed IR is shown in Figure 5.11 (BRAS Reference Scene 5, setup MS01 and MP01). The presented figures show both measurement and

[44] The class is part of the ITA-Toolbox for MATLAB®, `http://www.ita-toolbox.org`

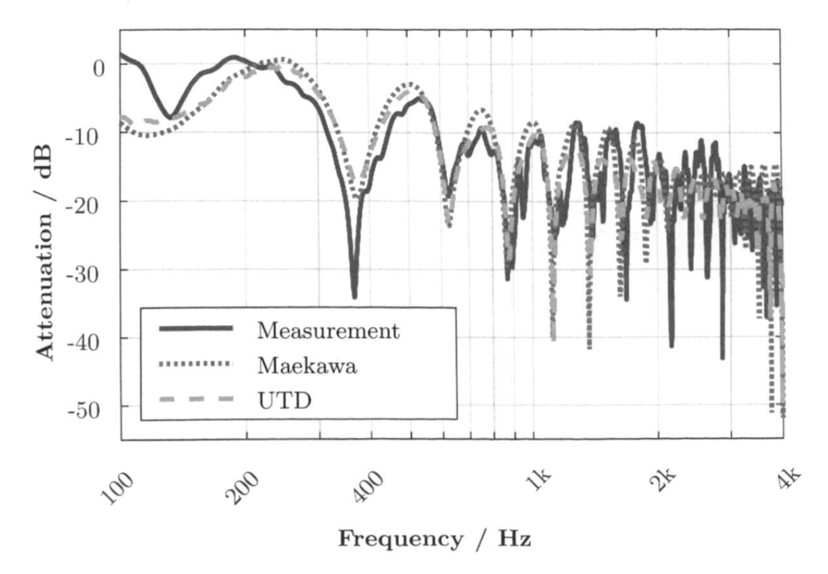

Figure 5.10: Measured and simulated TF of BRAS Reference Scene 5 (setup LS01 and MP01) [Asp+20] with manual insertion of transmission through the solid plate.

simulation referenced to the idealised free-field counterpart (omnidirectional point source). Reflection impedances have been considered perfectly reflecting ($\alpha = 0$). To cover wavefront divergence, spherical spreading loss is considered until the first occurrence of diffraction, and the combined spherical and cylindrical spreading loss of the diffracted wave is applied for the residual segments, as suggested by Kawai et al. [Kaw81]. The loudspeaker's on-axis TF is consequently applied to all emitted wavefronts. Additionally, each individual directional TF[45] is integrated depending on the emission angle (cf. also Figure 5.9). Finally, the on-axis loudspeaker directivity group delay has been reverted in order to match the temporal structure of the IR. Because the BRAS documentation does not specify a source sound power, the overall levels were estimated by comparing the pressure levels of a setup with the direct sound present (e.g., Scene 5, setup MP04 and LS04).

Based on the curves plotted in Figure 5.10 and Figure 5.11 it is stated that the approach combining the geometrical pathfinder and the acoustic modelling is able to reproduce the real-world acoustic environment adequately. The TFs exhibit

[45] The converted directivity dataset from the BRAS measurements have been normalised to the frontal direction

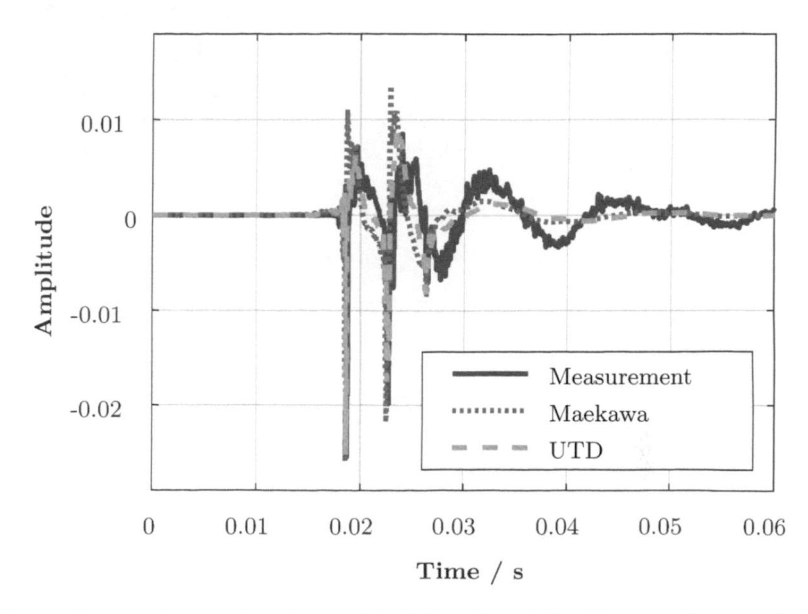

Figure 5.11: Measured and simulated IR of BRAS Reference Scene 5 (setup LS01 and MP01) [Asp+20] with manual insertion of transmission through the solid plate.

a matching comb filter structure with a reasonable proximity of the pressure amplitudes over frequency in the range between 300 Hz and 2 kHz. Areas with larger deviations are identified in the frequency range below 200 Hz, which can be explained by room modes of the anechoic chamber that are not sufficiently suppressed [Bri+21]. Also, with increasing wavelength, it is questionable if far-field conditions that are required for the validity of GA approaches are maintained in this scenario considering distances between the acoustic entities and surface boundaries. For example, in the measurement, the range of 2-3 kHz shows a distinct fine structure, which does not match that of the simulation. As the frequency increases, the observations drift into the discussion of uncertainties, since the heavily overlapping curves disallow a qualitative comparison. Last but not least, it is conceivable that the diffraction model, in particular the UTD for the diffraction at a screen, includes deviations from the real acoustic behaviour that require further investigation. Considering the temporal structure depicted by the IR of Figure 5.11, the timing and amplitudes of the first two impulses are matching accurately. However, the third impulse at 250 ms and the subsequent fluctuation is only indicated, but the amplitudes are not met. The similarities in the timing are essentially confirming that the geometrical propagation path lengths are well-aligned with the measured arrival times of the travelling wavefronts. It is noted, as each relevant simulated path includes a diffraction component and the individual contributions are *coherently superposed* at the receiver location, that the UTD diffraction model delivers consistent results that provide a reasonable overall sound pressure estimation here.

Listing 5.1: Obtaining a transfer function from a file of geometrical propagation paths exported by the pathfinder algorithm (MATLAB® code).

```
1  %% Geometrical propagation modelling
2
3  c = 344; % m/s
4  fs = 44100; % Hz
5  fft_degree = 14;
6  num_bins = 2^( fft_degree - 1 ) + 1;
7
8  % Instantiate class
9  gpsim = itaGeoPropagation( fs, num_bins );
10 gpsim.c = c;
11
12 % Load BRAS DAFF directivity, see also Appendix for conversion
13 gpsim.load_directivity( 'Genelec8020.v17.ir.daff', 'Genelec8020' );
14
15 % Load propagation paths (exported by pathfinder algorithm)
16 gpsim.load_paths( 'BRAS_scene5_LS01_MP01.json' );
17
18 dft_coeffs = gpsim.run(); % Magic happens here
19
20 % Plot using itaAudio
21 res = itaAudio();
22 res.samplingRate = fs;
23 res.fftDegree = fft_degree;
24 res.freqData = dft_coeffs;
25 res.plot_frequency
```

Altogether, by comparison with measurements regarding this specific scenario, it can be confirmed that the proposed approach describes a feasible method encouraging the employment of both the pathfinder algorithm and the acoustic modelling for geometrical propagation paths.

5.4 Pass-by auralisation

To verify the technical realisation of an auralisation framework for dynamic virtual environments, an evaluation based on the rendering result is required. Therefore,

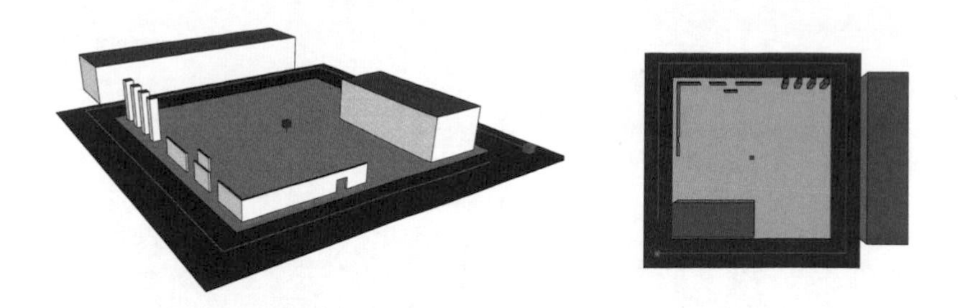

Figure 5.12: Auralisation test square with receiver in the middle and a dynamic sound source following the purple trajectory.

a virtual square model is sketched that consists of typical elements found in urban environments. Since a common use case of an urban noise situation is a pass-by event, the scenario contains a moving car that represents a dynamic noise source.

5.4.1 Scenario

The *auralisation square* is an artificial setting that is comparable to a city square with a circulating roadside. It is designed to test the quality of the pathfinder algorithm with partly complicated situations, the acoustic modelling accuracy regarding the continuity of the total sound field and the processing performance of the real-time auralisation system. The geometrical model of the site is depicted in Figure 5.12 and the receiver location is indicated by the central blue cube. The user is allowed to freely move inside the green lawn area, while the grey-coloured surrounding area is representing the street. The virtual car is following the purple trajectory counter-clockwise beginning at the location of the red cube. Several objects of different shape and dimension are placed between the lawn area and the street to provoke acoustically challenging constellations. The vehicle is elevated by 1 m, the receiver is located 1.7 m over ground and all building heights are above 2 m. In the vicinity of the first trajectory section, a large building is erected. The second section of the trajectory passes a rudimentary building structure with a flat rooftop that is placed on the other side of the road. The small-diameter pillars at the beginning of the third section represent a sequence of tree trunks. Three cascading screens follow in the next section, while the last section is mostly blocking the path between sound source and sound receiver by a long wall that includes a gateway-like gap in the middle.

5.4.2 Auralisation result

To evaluate the auralisation result, an offline routine is performed that sequentially executes the simulation, the modelling and the auralisation. In this way, intermediate results can be validated and the influence of different configurations can be investigated. A frame-by-frame simulation at a resolution of 86 Hz is performed with a maximum combined order of two, integrating paths containing up to two reflections and three diffractions. The acoustic modelling is calculated for the combined orders of one and two to investigate the influence of order-capping on the auralisation result. A UTD coefficient fade-out for paths of highest order that contain a diffraction as the final path segment has been applied to improve the sound field continuity, as discussed in Section 3.5. Per frame, a high-resolution sound transmission is calculated by coherently adding the complex-valued TFs of each individual path accounting for constructive and destructive interference for each frequency. Additionally, a set of parameters that is required by the auralisation DSP network is determined. The parameters include the energetic attenuation magnitude spectrum in third-octave resolution, a scalar gain value, which can also be negative, and the positive scalar propagation delay (cf. Table 4.2).

Figure 5.13 shows the A-weighted sound pressure level as perceived by an omnidirectional virtual microphone for a noise source emitting a flat sound power density spectrum at a level of $L_W = 105\,\text{dB}$ that moves along the indicated trajectory at a velocity of $30\,km/h$. The solid blue curves are relating to the left blue axis and present the levels of the superposed TFs integrating paths up to combined order one (light blue) and combined order two (dark blue), respectively. The dotted red curve depicts the sound pressure level results based on the auralisation DSP parameters, as will be discussed, see Section 5.4.3. The solid green curve is related to the right green axis and shows the total number of propagation paths per frame. Additionally, the particular distribution of reflections and diffractions among the paths is indicated by the light green and light yellow areas below the total paths curve. A visualisation of geometrical propagation paths for four representative source locations is shown in Figure 5.14.

All curves exhibit the expected influence of the geometrical environment on the transmitted sound pressure levels. At the beginning, after $\sim 1\,\text{s}$, the source is located behind the large building and only diffracted paths of second order establish a valid connection to the receiver, since the first-order curve is interrupted in this region. Then, the reflection at the facade of the second section becomes

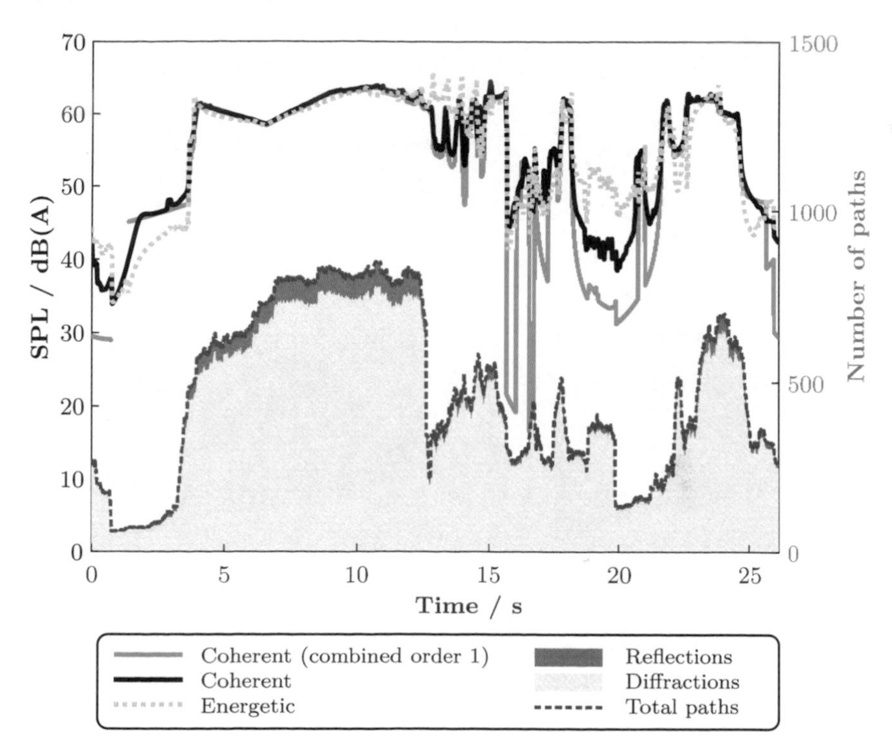

Figure 5.13: A-weighted sound pressure levels as perceived by the receiver for a noise source of $L_W = 105\,\mathrm{dB}$ and composition of propagation paths taking one turn around the test square of Figure 5.12 at a speed of $30\,km/h$.

audible and the sound pressure level rises constantly until the direct sound path instantaneously appears at $\sim 4\,\mathrm{s}$. Here, the direct sound and the reflected paths dominate the sound field until the source reaches the opposite corner of the square at $\sim 12\,\mathrm{s}$. In the following, a complicated fluctuation is depicted that is related to the rapidly appearing and disappearing direct sound path behind the thin pillars that represent the tree trunks. The moment the source enters the shadow zone behind the screen, approximately at the centre of the third section after 16 s, the sound pressure level drops immediately. The influence of the cascaded screens afterwards, in range of $16\,\mathrm{s} - 18\,\mathrm{s}$, is indicated by slightly fluctuating values at a comparable average level, which is only interrupted by the slit opening towards the end of the third section where the direct sound appears for a short instance. Along the final trajectory beginning at $\sim 20\,\mathrm{s}$, an average continuous increase

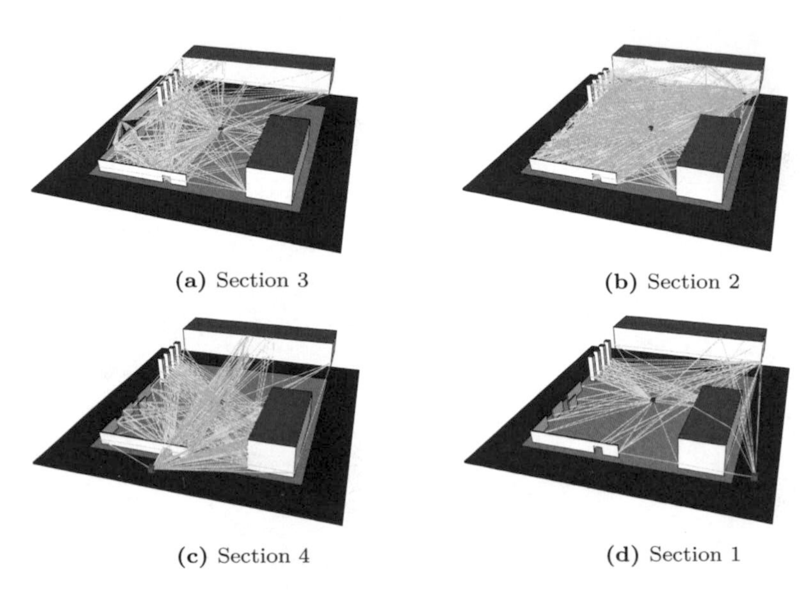

(a) Section 3

(b) Section 2

(c) Section 4

(d) Section 1

Figure 5.14: Visualisation of propagation paths (combined order 2) for particularly interesting source locations of each section in the auralisation square scenario.

is observed with two short peaks at $\sim 21\,\mathrm{s}$ and $\sim 22\,\mathrm{s}$ that are related to the opening of the long wall, when reflected sound and direct sound appears and disappears. Between $\sim 23\,\mathrm{s}$ and $\sim 24\,\mathrm{s}$, a section with both direct sound and a strong reflection from the opposite side of the square is shown, that disappears towards the end. After the direct sound is finally blocked by the large building at $\sim 25\,\mathrm{s}$ again, the sound pressure levels drop rapidly and approach the initial values at $0\,\mathrm{s}$.

Considering the progression of the dark blue curve representing the sound pressure levels calculated with the paths from the combined second-order simulation and using a coherent superposition of the TFs, the results largely reflect the expected sound pressure levels at the receiver location. In particular, a continuous development is depicted, since the curve neither contains gaps nor steps. Such undesired effects are present in the combined first-order simulation that obviously is incapable to determine all required propagation paths to maintain a continuous sound field. However, in areas with many reflected propagation paths, theses effects appear to be masked as the two curves are very similar. Hence, to adequately auralise challenging situations such as a large building at the beginning or the cascaded screens of the third section, higher orders must be considered. The significance to chose an adequate combined reflection and diffraction order for the given context is more evident at the gap in the wall of the last section, which is of comparable quality for the instance when the direct sound appears at $\sim 22\,\mathrm{s}$, but reveals highly deviating sound pressure levels for the event where strong reflections are involved at $\sim 21\,\mathrm{s}$.

The geometrical propagation paths shown in Figure 5.14 are visualising four representative situations, one of each section. In the bottom-right image, the source is located behind the large building and valid paths include at least one diffraction. In the next section, as depicted in the top-right visualisation, the source is passing the reflecting facade and consequently an extensive number of paths is exhibited consisting of all possible combinations of reflection and diffraction orders. In the situation of the top-left image, the source has disappeared behind the screens, which corresponds to the high level drop of Figure 5.13 at $\sim 17\,\mathrm{s}$. In the last section, as shown by the bottom-left image, the sound source is directly visibly to the receiver through the opening of the long wall, which is related to the level increase at $\sim 22\,\mathrm{s}$.

5.4.3 DSP coefficient modelling

Since the concept of the presented real-time auralisation module employs a dedicated DSP network for dynamic outdoor environments, a specific set of parameters is required to control the individual units, see Section 4.7.2. In particular, the short IIR filter units approximate the frequency response with respect to the absolute magnitude, however, are not able to maintain phase information. In contrast, adding complex-valued TFs delivers the desired result, as contributions are added coherently. Therefore, the combination of individual attenuation spectra produces a result that is related to an energetic summation. Regarding the sound field continuity property, a problematic increase of energy is observed at the transition from the shadow region into the illuminated region if both the direct sound and the diffracted sound field component are added by their absolute magnitude spectra. According to the UTD model, the diffraction contribution decreases the total sound field by wavelength-dependent interference at complex-valued magnitudes that are almost equal to the direct sound level. Hence, the compensating property of the UTD is lost. Nonetheless, a concept for improvement at the corresponding boundaries has been integrated. The approach makes use of the fact that the complex-valued UTD coefficients are aligning their phases at the critical boundaries to deliver a continuous development for each frequency. Hence, the filtered signals are *subtracted* and not added in certain diffraction zones in order to avoid the undesired effect of overly increased levels, as both signals are approximately in-phase. Therefore, the sign of the *gain* is modified accordingly, which results in phase inversion of the path contribution in the DSP network where necessary. Although the approach does not accurately maintain continuity for all frequencies if the phases are not precisely aligned, a substantial improvement of the overall sound pressure level is achieved. Figure 5.13 in the previous section depicts the energetic superposition of the individual propagation paths as red dotted line considering a combined order of two (including third-order diffractions with fade-out function, see Section 3.5). The phase-inverted addition of signals is integrated and can be observed, for example, at $\sim 4\,\mathrm{s}$ where the direct sound appears and the sound pressure level based on the auralisation parameters does not surpass the level from the coherent calculations significantly. However, considering the region where the source moves behind the pillars between $\sim 13\,\mathrm{s}$ and $\sim 15\,\mathrm{s}$, the undesired effect is observed. Additionally, the levels are largely overestimated between $\sim 18\,\mathrm{s}$ and $\sim 20\,\mathrm{s}$, which appears to be a result of an energetic summation of many small diffraction paths. The same holds for the initial phase behind the large building at $\sim 2\,\mathrm{s}$, where the accumulation of several hundred paths leads to an overestimated sound pressure level and a underestimation when the source proceeds towards the location where the direct sound appears at $\sim 4\,\mathrm{s}$. Apart from these regions that require further investigation, the sound pressure levels based on the auralisation parameters are sufficiently similar to the coherent high-resolution curve and do not reveal unnatural onsets or steep declines.

The presented auralisation pipeline employs autoregressive IIR filter units of user-configurable order and the performance of the frequency-shaping is shown in Figure 5.15. A representative attenuation spectrum is depicted that is commonly found for individual propagation paths with low-order diffraction components. The blue spectrum shows a representative target spectrum as delivered by the acoustic propagation modelling and the dotted red curve represents a reference implementation using a 128-taps FIR filterbank. It can be seen that the low-frequency range between $\sim 20\,\mathrm{Hz}$ and $\sim 80\,\mathrm{Hz}$ is underestimated and the region around $\sim 200\,\mathrm{Hz}$ is slightly overestimating the target spectrum, but the curves are otherwise in agreement. The dashed lines are representing the IIR counterparts for orders 4, 10 and 40, for which the autoregressive coefficients are determined by the Burg scheme [Bur68; De +96]. The results suffer from a similar deviation in the low-frequency region and also exhibit a range between $\sim 400\,\mathrm{Hz}$ and $\sim 4\,\mathrm{kHz}$

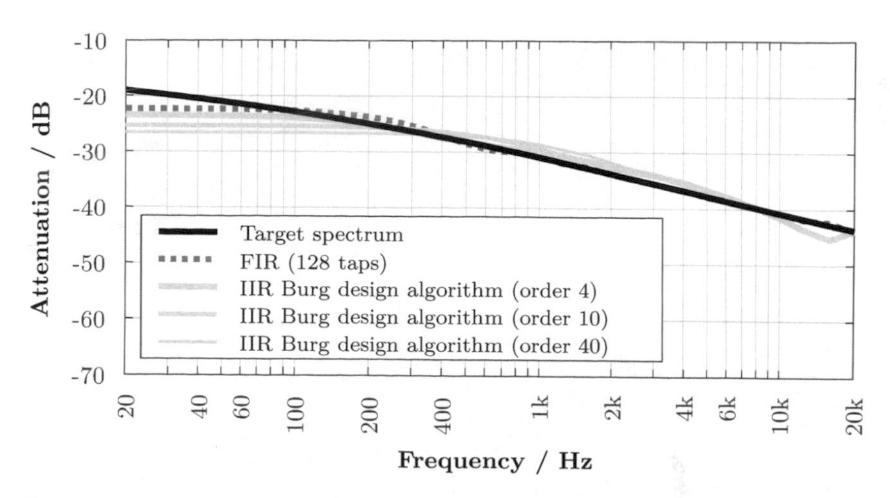

Figure 5.15: Quality evaluation of the IIR filter using a Burg scheme for a typical urban attenuation spectrum with a diffraction component [Bur68; De +96] (cf. also Section 4.7.2).

where attenuation levels are increased by several dB, at least for the lowest filter order. Between ~ 10 kHz and ~ 20 kHz, another underestimation can be seen.

In the context of real-time auralisation, the incorporation of IIR filter units with a Burg algorithm present a good compromise between accuracy regarding the target spectrum and computational efficiency, since filter orders of four are already delivering results that are in good accordance with the desired attenuation spectrum. It should be noted, that the IIR filter units are not used for more diverse spectra, since each individual outdoor propagation path attenuation spectrum typically has a low-pass character due to diffractions and air absorption, see Section 3.4. Consequently, the result of Figure 5.15 supports the feasibility of low-order IIR filter units in combination with the Burg design approach to account for frequency-dependent attenuation for the real-time auralisation of outdoor sound propagation.

5.5 Real-time VR example

The verification of the presented real-time auralisation approach for virtual outdoor environments is concluded by the description of an example application

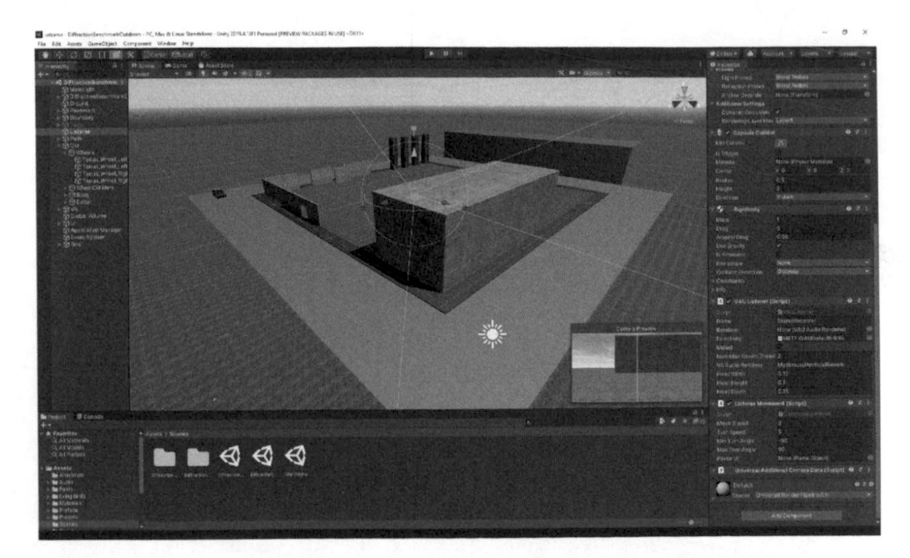

Figure 5.16: Virtual reality design application *Unity* © showing the auralisation demonstration scene with a scripted moving car taking turns around the square. An interactive user can move freely inside the square area using a keyboard and a mouse or a VR hardware setup.

that is running on a desktop computer and uses interactive VR technology. The demonstration system is based on the VR designer *Unity*®[46], which provides visual rendering and the interfacing to input devices and displays. The scene can be either configured for a desktop environment using a regular screen with a mouse and a keyboard or can be configured for a VR setup, for example, using an HMD with interactive controllers that are tracked in real-time. In both cases, the auralisation output stream is reproduced by headphones, as it contains binaural cues.

Figure 5.16 shows the urban scenario inside the VR designer application. As in the previous sections, the auralisation square is used and the dynamic sound source, represented by the car, is following the track according to Figure 5.12. While the simulation scheduler of the auralisation engine is using the same geometrical model, the faces of the visual model have been assigned with textures to achieve a more realistic impression. Apart from that, the visual and acoustic models are exactly congruent in the world coordinates of the virtual environment.

[46] Unity Technologies, `http://www.unity3d.com`

Regarding the acoustic modality, the native audio engine of the VR application is deactivated and the auralisation is performed by VA[47] running as a stand-alone server application. To transmit dynamic motion information, a set of VA classes are available that communicate with the auralisation instance by a network connection. Each acoustic entity of the virtual environment is extended by a corresponding class, for instance, the car is assigned the sound source class and the virtual camera is connected to the receiver class. If these objects are moved, for example, by tracking from the VR setup, the motion streams are automatically forwarded to the auralisation engine triggering a scene update, see Section 4.4.2 and Section 4.7.

The real-time auralisation system uses an ASIO[48]-compatible high-end audio interface with a buffer size of 128 samples at a sampling rate of 44.1 kHz resulting in an audio processing time budget of 1.2 ms.[49] A rendering module of class *BinauralOutdoorNoise* is activated in the configuration file of VA and the binaural output is routed to a reproduction module that forwards the stream directly to the headphones. As per default, the rendering engine uses six PDs with a convolution filter length of 256 samples for the HRIRs in the spatialisation unit, see Section 4.7.2. The interpolation method in the combined FD and Doppler effect component is set to linear interpolation and the IIR filter units operate at an order of four. Furthermore, a simulation scheduler is instantiated that initialises one worker back-end executed in a parallel thread on the same machine. A translation filter to skip motion below 10 cm is applied and a rate filter at 30 Hz is activated to synchronise the simulation results at a moderate constant update rate. To limit the simulation runtime and assure a sufficiently high update rate, a combined order of one is chosen.[50]

During execution on a desktop PC[51], the simulation runtime and the performance of the DSP network was recorded. The values can be obtained from Table 5.1 and show that the loading of the environment including the geometry pre-processing requires 41.0 ms in the initialisation phase. Subsequent simulations are then ex-

[47] Virtual Acoustics (VA), `http://www.virtualacoustics.org`
[48] ASIO™ is an audio driver architecture by Steinberg®, `http://www.steinberg.de`
[49] The buffer size and sampling rate is configurable by the user.
[50] This configuration is undesirable but necessary for real-time rates in a VR system that hosts the simulation scheduler and the auralisation rendering on the same machine. However, it is in contradiction to the findings in the previous sections and hence, the auralisation result suffers from the flaws of an insufficient simulation, as discussed in Section 5.4.2. Distributing the simulation load to dedicated machines is a vital option to provide more computation load for the auralisation and increase the simulation order.
[51] Microsoft® Windows™ 10 64-Bit operating system, Intel® Core™ i7-10700K CPU @ 3.80 GHz, 32.0 GB DDR3 memory, NVIDIA® GeForce™ GTX 1050 Ti graphics card, Steinberg® UR22C™ 32-bit audio interface

Table 5.1: Statistical evaluation of the runtimes in the auralisation engine.

Component	Average	Std.-Dev.	Cycles
Environment initialisation	41.0 ms	0.54 ms	100
Simulation	20.1 µs	5.73 µs	275
Audio processing	0.56 ms	0.18 ms	$\sim 50k$
Wavefront total	5.45 µs	2.43 µs	$\sim 3.9M$
Wavefront VDL read	0.43 µs	0.31 µs	$\sim 3.9M$
Wavefront IIR filtering	4.48 µs	2.29 µs	$\sim 3.9M$
Wavefront interpolation	0.16 µs	3.05 µs	$\sim 3.9M$

ecuted at an average of 20.1 µs due to the low order. With a time consumption of 0.56 ms for the audio processing, the DSP network uses approximately half of the available time budget of 1.2 ms and integrates an average of 90 individual wavefronts. Considering a single wavefront, the total amount of 5.45 µs is required to produce an individual contribution. Regarding the single units that are performed per wavefront separately, the VDL read-out routine providing the corresponding variable-length buffer uses 7.7% and the subsequent fourth-order IIR filtering unit requires 82% of the accumulated processing time, while the interpolation only needs 2.9%.

Based on these results, the combination of the pathfinder algorithm and the DSP network is generally applicable for a real-time VR environment, since the delivered simulation update rate is sufficient and the audio processor is not overloading. As the DSP network gives some headroom considering the current configuration, either a quality improvement can be achieved or the total number of processed propagation paths can be scaled. Regarding the quality, for example, the number of PDs in the binaural spatialisation module can be increased to provide a higher directional HRIR sampling. Additionally, a spline interpolation provides a higher SNR, which is particularly desirable for tonal signals as they do not mask the unwanted overtones of the linear interpolation. An increased quantity of propagation paths of up to 200 individual wavefronts is conceivable for this setup. For instance, considering rapid toggling between two individual locations, a second receiver can be added. This is particularly important for portal navigation in a VR setup, since the user is appearing at a distant location, which corresponds to an unphysical motion that results in audible artefacts. In contrast, if the target location is accounted for by a separate receiver, the output can be instantaneously switched delivering a smooth auralisation. Similarly, a second sound source can be included, for example, by representing the car with two separate sources, one for the engine compartment and one for the exhaust. Naturally, another car can be added as well or a stationary sound source like a water fountain can be placed inside the square to investigate, for instance, the psycho-acoustic masking effect.

Since the representation of first-order paths is insufficient for some source-receiver constellations in this scenario, as depicted in Figure 5.13, it is desirable to include further propagation paths of higher order. However, because the implementation does not apply a priority ranking, an increased simulation order leads to an overload of the audio processor as the expected number of paths exceeds the processable individual wavefronts and the real-time capability is lost. Approaches to overcome this issue are discussed in the outlook, see Section 7.

6 Conclusion

In this thesis, a comprehensive auralisation system has been implemented with the aim to auralise dynamic outdoor sound propagation at real-time rates. The specific realisation is based on decisions that were guided by the careful evaluation of conceivable concepts and established methods, which will be critically reviewed in the following. Additionally, implications on the employment for future projects and possible practical issues are discussed.

Performing an auralisation of virtual environments by assuming quasi-stationary time frames and subsequently cross-fade results is incompatible with fast-moving sound sources. Hence, it does not represent a valid option in the context dynamic urban environments. This aspect has supported the idea to develop a DSP network that processes individual paths by a SIMO-VDL to realise the time-variant propagation delay in a physically adequate way and to integrate the Doppler shift for each individual wavefront. As a consequence, a class of filtering must be included for each propagation path to account for frequency-dependent attenuation, which imposes a high computational burden. To achieve real-time update rates, the number of polygons representing the scene and/or the number of sound sources must be considerably limited, which stands in contradiction to realistic urban environments, as they are commonly large in their extent and contain numerous sound sources. This inconvenient relationship must be acknowledged and is identified as the main reason why the proposed system, in its current stage, is not feasible for highly complex scenarios that require the simultaneous consideration of several hundred propagation paths under real-time constraints. Since the employed DSP units are reasonably optimised, the only remaining possibility to increase the rendering capacity is the distribution of DSP calculations, for instance, to a dedicated hardware processing unit. However, another promising approach is mentioned in Section 4.4.2 and discussed in Section 5.2.2. It deals with the issue of ranking paths by an importance metric with the goal to promote most relevant contributions in case the desired number of propagation paths is exceeding the processing capacity of the DSP network. An improvement is achieved by either providing an auralisation that is perceptually indistinguishable from a result using a higher DSP capacity or by increasing the general plausibility, for example, in an interactive VR environment. The determination of such an importance metric is subject to future investigation and may involve, for example, the employment of psycho-acoustic methods, like loudness evaluation and spatio-temporal masking, by considering the propagation paths for each source separately or by integrating the contributions of all sound sources at the receiver location.

A similar issue is evoked by the pathfinder algorithm, which determines an exponentially growing number of propagation paths with respect to the reflection and diffraction order. The algorithm has been realised to find all possible propagation paths from a source to a receiver location with certainty in order to conceptually avoid the problem of disregarding relevant paths, as conceivable, for example, by spatial undersampling in a ray tracing approach. However, the extensive number of paths already found in a moderately complex scene demands a more sophisticated abortion criterion that ideally proves general validity and complements or extends the approaches described in Section 3.6.8. Additionally, the presented pathfinder algorithms is incapable to determine propagation paths via inconveniently shaped openings, which leads to missing sound field components that can be an issue in specific situations (cf. Section 3.6.3). In the context of auralisation, however, this misconception is reasonable if the UTD method is employed, since it is not able to solve diffraction at finite wedges in the first place.

Another noteworthy issue that requires a critical review is revealed by the figures depicted in Section 5.4.2 and discussed in Section 5.4.3. It is concerned with the transformation of propagation modelling results into DSP coefficients, as required by the auralisation pipeline. According to the sound pressure levels shown in Figure 5.13, a direct comparison with the results based on the high-resolution TFs is likely to reveal a noticeable difference. It must be noted, however, that the solution using TFs is only a theoretical concept that is determined for this particular plot, as it assumes a frame-by-frame stationary situation and is not compatible with moving sound sources. Hence, it remains an open question if the auralisation result using the DSP coefficients gains plausibility if the overall sound pressure values can be approximated more accurately. The circumstance that only a gain value and an energetic spectrum is available to represent the contribution of each individual path in the DSP network is challenging. In the current state, the sign of the gain is flipped for each path separately to approximate the characteristic destructive interference of the UTD coefficients at critical boundaries, delivering a general improvement that is yet vulnerable to inconvenient constellations. To further decrease the deviation in the auralisation result, a clever solution is required that may be found, for example, by taking all propagation paths from the source to the receiver into account, since they are ultimately superposed in the rendering pipeline.

Finally, the binaural spatialisation module of the auralisation engine is designed to limit the computational resources that must be allocated for the HRIR convolutions. However, for a plausible auralisation, a clear conception how to determine a sufficient number of PDs and appropriately configure the angular distance metric in the clustering module result remains open. It is conceivable that a listening experiment investigating the performance of the clustering engine with respect to the available computational resources can formulate an optimum prediction of the parameters for a given number of incident wavefronts. However, the perceptual evaluation is likely to include a contextual dependency and varies, for example, if either many paths are considered that are emitted from a single sound source or many different sound sources are clustered under free-field conditions. For this reason, it appears more reasonable to recommend the determination of the appropriate configuration individually by a preceding informal listening test.

The evaluation of the presented auralisation approach is separately investigated in Chapter 5 and the results are largely supporting the successful realisation and the applicability of the framework. The possibility to perform acoustic simulations incorporating highly-diffracted sound using relatively fast GA methods opens the door to find reasonable simplifications. For example, an importance ranking method regarding different aspects like transmitted energy is presented that effectively decreases the excessive path count by concentrating on relevant contributions. Furthermore, it is confirmed that the propagation modelling, which uses the geometrical propagation path results delivered by the pathfinder algorithm, is in good agreement with the measurement of a relatively simple scene. The combination of the simulation and modelling methods delivers propagation results, such as coherent TFs and approximate attenuation band spectra, that largely reproduce the expected sound pressure levels at a receiver location of a dynamic test scenario. For instance, occlusion effects are shown when the source is located behind a large building and a short level increase is observed if the source passes an opening of a noise barrier. Performing the same scenario in a VR environment, however, is only possible if profound simplifications regarding the simulation engine are applied that is predominantly required to maintain a limited path count in order to avoid an overload in the real-time audio processor.

To conclude, the presented approach is capable to provide an auralisation result that is highly relevant for future investigations concerning noise topics in urban environments. In particular, if an offline evaluation is desired, sequential high-resolution simulations producing frame-by-frame TFs can be employed at the cost of a prolonged calculation time. In contrast, if the described method is complementing a VR application, the possibilities regarding the complexity of the virtual environment and particularly the number of sound sources is limited in favour of real-time capability. If the system shall gain general applicability, further acceleration concepts and simplifications in the DSP network are required to maintain real-time capability for more complex scenarios, in particular, a higher number of dynamic sound sources must be provided. As suggested, the auralisation routine must realise a complex priority filtering stage in order to deliver a constant path count that complies with real-time constraints imposed by the DSP network.

7 Summary and Outlook

A real-time auralisation framework has been realised that is able to acoustically render virtual outdoor environments with dynamic, fast-moving sources and interactive receivers in a static built environment. The concept is founded on three major innovations, namely

1. a pathfinder algorithm determining geometrical propagation paths with arbitrary reflections and diffractions,
2. an acoustic propagation modelling using the UTD method for continuous diffraction handling and
3. a DSP network layout that is specifically designed for dynamic, time-variant auralisation at real-time rates.

The implementation of the auralisation approach is motivated by the desire to present an acoustic context to a group of non-experts in an easily comprehensible way in order to complement common noise assessment methods. The demonstration of an outdoor noise scenario that is intuitively perceivable without expertise, as provided by an interactive VR environment, requires a physics-based concept that can reproduce fast-moving sound sources and dynamic receivers in an urban setting in a plausible fashion. Hence, a fast method to determine characteristic acoustic propagation features such as reflections at surfaces, diffractions at edges and the Doppler effect is required. The calculation of sound propagation is primarily based on the classical models, as described in Chapter 2. However, regarding the available acoustic diffraction models, the UTD method is of particular interest as it delivers a sufficiently accurate attenuation spectrum and provides sound field continuity across critical boundaries where the direct and reflected sound field abruptly appears and disappears, which is deemed equally important (cf. Section 2.3.2). Furthermore, a *pathfinder* algorithm of the GA class is presented in Section 3.6 that deterministically finds geometrical propagation paths in a static polygonal environment from a source to a receiver location, while integrating arbitrary combinations of reflections and diffractions. Because the approach delivers individual geometrical paths, it is especially appropriate for outdoor propagation models that apply time-variant effects like the Doppler shift for each wavefront separately, see Section 3.4.7. This feature is relevant in the rendering stage of the framework that combines the simulation and modelling results in order to generate a plausible, artefact-free auralisation result. Instead of superposing propagation contributions to a single IR and employing a convolution per source-receiver-pair, the presented DSP layout consequently processes individual wavefronts based on an auralisation timeline concept, described in section 4.3.1, to a) adequately integrate the dynamic nature of outdoor environments and b) avoid cross-fading artefacts, see Section 4.6.2 and Section 4.6.3. Therefore, a time-variant rendering pipeline is presented in Section 4.7 consisting of a central SIMO-VDL per source-receiver pair, short IIR filter units that apply individual propagation attenuation per incident wavefront and an interpolation routine that renders the Doppler shift. A final spatialisation module incorporates binaural cues using a directional clustering scheme in order to group incoming wavefronts at the receiver location and route them through a fixed number of PDs, each performing a HRIR convolution. Consequently, a manageable computational load is maintained that does not scale with the number of propagation paths. Furthermore, an approach is integrated that accounts for azimuthal deviations of each incoming wavefront with respect to the assigned PD based on an ITD model adjusting the FD to match the inter-aural delay difference, see Section 4.7.2. To verify the general applicability, the simulation algorithm is investigated in Section 5.2 considering a typical street canyon scenario. It is shown that paths with multiple diffractions are highly attenuated and are largely irrelevant given the dominance of reflections based on a priority ranking by transmitted energy. The feasibility of the propagation modelling is discussed in Section 5.3 by comparing measurement and simulation of a scenario consisting of a screen placed on a reflecting plane that requires the integration of a diffraction model. The auralisation capability of the presented system is evaluated in Section 5.4.2 by a virtual square scenario with a circulating, dynamic car. Different sections with challenging diffraction constellations regarding the reproduction of a continuous sound field are evaluated by the simulated sound pressure levels at a central receiver location, exhibiting general agreement with the expectations as all attenuation effects can be identified. However, the necessity to simulate a certain order of reflections and diffractions for this scenario is evident, as otherwise unrealistic gaps with total silence occur and the continuity of the sound pressure levels over time is deteriorated. Additionally, regarding the results based on the coherent TFs, a difference of the auralisation output employing the real-time DSP network is depicted. The superposition of individual wavefronts applying an IIR filter cannot add sound pressures with the corresponding phases accurately, which is prominent in critical regions where diffractions dominate the total sound field in the absence of reflections. Finally, in Section 5.5, a virtual scenario representing an urban setting is described that employs the presented real-time auralisation system and can be experienced in an audio-video VR application using standard hardware. To comply with real-time constraints, the combined order of reflections and diffractions is limited to one, since a maximum capacity of up to 200 individual sound paths must be maintained.

The presented auralisation framework realises a promising foundation for future research in the context of urban noise assessment. However, several aspects are identified that require further improvement.

As characteristic for deterministic GA algorithms, the IEM approach exhibits an exponential growth rate with order of reflections and diffractions, which makes high-order simulations infeasible considering the execution time and the memory consumption. Additionally, the presented implementation creates a visibility matrix that grows quadratically with the number of faces and edges of the geometry mesh, which further decreases the applicability. A more efficient data structure, for example, realising a binary space partitioning, is a viable solution to significantly improve the runtime enabling both the simulation of more complex scenarios and more source and receiver entities in parallel.

Furthermore, the consequence that certain situations with inconveniently-shaped openings are not detectable by the pathfinder algorithm must be addressed in the future, if an environment is likely to contain such constellations. In this case, a solution that incorporates finite edge diffraction must be realised. For example, the pathfinder can be extended to additionally provide propagation paths via the vertices of the polygonal mesh. A subsequent modelling must be based on the BTMS model, which substantially increases the complexity of both the simulation and the propagation modelling.

The results of the pass-by auralisation demonstrate that the approach is highly relevant for the acoustic evaluation of dynamic urban environments. Especially, the possibility to simulate sound pressure levels by coherent superposition of individual propagation paths of higher orders is a promising feature to extend current noise prediction methods, since it delivers a high temporal and spatial resolution and deterministically integrates the contribution of diffracted sound. Nonetheless, the excessive number of propagation paths that are delivered by the simulation method must be effectively limited in order to maintain a manageable size. As indicated, a novel priority scheme must be realised to separate irrelevant contributions and dismiss negligible paths. Because limiting the order of reflections and diffractions is not sufficient for certain situations, more sophisticated abortion criteria must be evaluated that are motivated by technical and perceptual aspects. As a future investigation, it is conceivable to produce various auralisation results of a given scenario under different conditions and perform a comparative listening experiment. For example, it is of great interest to find a perception threshold of minimum propagation paths that must be included by selecting a set of high-ranked propagation paths considering the estimated transmitted energy. However, more elaborate approaches must be investigated for highly complex environments with many sound sources. In particular, applying psycho-acoustic models predicting spectral and temporal masking of individual paths or sound sources in general must be investigated in the future, if it is desired to auralise many-source environments under real-time constraints. This way, for example, it is possible to increase the simulation order of the presented VR scenario and add further sound sources without overloading the audio processing, if the path filtering routine is configured to deliver the 200 most important propagation paths. Because the DSP network is already highly efficient and further simplifications of the employed units are inconceivable, the invention and validation of a perception-related priority concept to limit the determined propagation paths is considered the most promising aspect to improve the presented framework for the real-time auralisation of outdoor sound propagation in the future.

Bibliography

[AB79] J. B. Allen et al. "Image method for efficiently simulating small-room acoustics". In: *J. Acoust. Soc. Am.* 65.4 (1979), pp. 943–950. DOI: 10.1121/1.382599.

[Ant+10] L. Antani et al. *Fast geometric sound propagation with finite edge diffraction.* Tech. rep. Technical Report TR10-011, University of North Carolina at Chapel Hill, 2010., 2010.

[Ant+12] L. Antani et al. "Efficient finite-edge diffraction using conservative from-region visibility". In: *Applied Acoustics* 73.3 (2012), pp. 218–233. DOI: 10.1016/j.apacoust.2011.09.004.

[AV14a] M. Aretz et al. "Combined wave and ray based room acoustic simulations of audio systems in car passenger compartments, Part I: Boundary and source data". In: *Applied Acoustics* 76.0 (2014), pp. 82–99. DOI: 10.1016/j.apacoust.2013.07.021.

[AV14b] M. Aretz et al. "Combined wave and ray based room acoustic simulations of audio systems in car passenger compartments, Part II: Comparison of simulations and measurements". In: *Applied Acoustics* 76.0 (2014), pp. 52–65. DOI: 10.1016/j.apacoust.2013.07.020.

[AS10] A. Asheim et al. "Efficient evaluation of edge diffraction integrals using the numerical method of steepest descent". In: *J. Acoust. Soc. Am.* 128.4 (2010), pp. 1590–1597. DOI: 10.1121/1.3479545.

[AP13] A. Asheim et al. "An integral equation formulation for the diffraction from convex plates and polyhedra". In: *J. Acoust. Soc. Am.* 133.6 (2013), pp. 3681–3691. DOI: 10.1121/1.4802654.

[APV14] L. Aspöck et al. "A Real-Time Auralization Plugin for Architectural Design and Education". In: *EAA Joint Symposium on Auralization and Ambisonics (2014), Berlin, Germany.* 2014, pp. 156–161.

[Asp+18] L. Aspöck et al. "Application of virtual acoustic environments in the scope of auditory research". In: *XXVIII Encontro da Sociedade Brasileira de Acústica : SOBRAC 2018.* XXVIII Encontro da Sociedade Brasileira de Acústica, Porto Alegre (Brazil). 2018. DOI: 10.17648/sobrac-87162.

[Asp+19] L. Aspöck et al. "Auralization of interactive virtual scenes containing numerous sound sources". In: *Proceedings of the 23rd International Congress on Acoustics : ICA 2019.* 23rd International Congress on Acoustics : integrating 4th EAA Euroregio 2019, Aachen (Germany). Deutsche Gesellschaft für Akustik e.V. (DEGA), 2019, pp. 6401–6401. DOI: 10.18154/RWTH-CONV-239281.

[Asp+20] L. Aspöck et al. *BRAS - Benchmark for Room Acoustical Simulation.* 2020. DOI: 10.14279/depositonce-6726.3.

[Beg00] D. Begault. *3-D Sound for Virtual Reality and Multimedia.* 2000.

[BVV93] A. J. Berkhout et al. "Acoustic control by wave field synthesis". In: *The Journal of the Acoustical Society of America* 93.5 (1993), pp. 2764–2778.

[BD09] F. Besnard et al. *Road noise prediction 2 - Noise propagation computation methodincluding meteorological effects (NMPB2008).* 2009.

[BT57] M. A. Biot et al. "Formulation of wave propagation in infinite media by normal coordinates with an application to diffraction". In: *J. Acoust. Soc. Am.* 29.3 (1957), pp. 381–391.

[Bla97] J. Blauert. *Spatial hearing: the psychophysics of human sound localization.* MIT press, 1997.

[BJ97] J. Blauert et al. "Sound-quality evaluation–a multi-layered problem". In: *Acta acustica united with acustica* 83.5 (1997), pp. 747–753.

[Bla+00] J. Blauert et al. "An interactive virtual-environment generator for psychoacoustic research. I: Architecture and implementation". In: *Acta Acustica united with Acustica* 86.1 (2000), pp. 94–102.

[Bor84] J. Borish. "Extension of the image model to arbitrary polyhedra". In: *The Journal of the Acoustical Society of America* 75.6 (1984), pp. 1827–1836. DOI: 10.1121/1.390983.

[BW97] M. Born et al. *Principles of Optics: Electromagnetic Theory of Propagation, Interference and Diffraction of Light.* Cambridge University Press, 1997.

[Bot+02] M. Botsch et al. *Openmesh - a generic and efficient polygon mesh data structure.* 2002.

[BM78] J. Bremhorst et al. "Impulse wave diffraction by rigid wedges and plates". In: *J. Acoust. Soc. Am.* 64.S1 (1978), S64–S64. DOI: 10.1121/1.2004312.

[Bre78] J. H. Bremhorst. "Impulse wave diffraction by rigid wedges and plates". PhD thesis. Monterey, California. Naval Postgraduate School, 1978.

[Bri+21] F. Brinkmann et al. "A benchmark for room acoustical simulation. Concept and database". In: *Applied Acoustics* 176 (2021), p. 107867.

[BKG11] A. Brown et al. "Towards standardization in soundscape preference assessment". In: *Applied Acoustics* 72.6 (2011), pp. 387–392.

[Bur68] J. P. Burg. "A new analysis technique for time series data". In: *Paper presented at NATO Advanced Study Institute on Signal Processing, Enschede, Netherlands, 1968* (1968).

[CSF05] P. T. Calamia et al. "Integration of edge-diffraction calculations and geometrical-acoustics modeling". In: *Proceedings of forum acusticum.* 2005.

[CS07] P. T. Calamia et al. "Fast Time-domain Edge-diffraction Calculations for Interactive Acoustic Simulations". In: *EURASIP J. Appl. Signal Process.* 2007.1 (2007), pp. 186–186. DOI: 10.1155/2007/63560.

[CMS08] P. T. Calamia et al. "Diffraction Culling for Virtual-Acoustic Simulations". In: *Acta Acustica united with Acustica* 94.6 (2008), pp. 907–920. DOI: 10.3813/AAA.918108.

[Cal09] P. T. Calamia. "Advances in edge-diffraction modeling for virtual-acoustic simulations". PhD thesis. Princeton University, 2009.

[CFP15] A. Can et al. "Accounting for the effect of diffuse reflections and fittings within street canyons, on the sound propagation predicted by ray tracing codes". In: *Applied Acoustics* 96 (2015), pp. 83–93. DOI: 10.1016/j.apacoust.2015.03.013.

[CK88] C. S. Clay et al. "Numerical computations of time-domain diffractions from wedges and reflections from facets". In: *J. Acoust. Soc. Am.* 83.6 (1988), pp. 2126–2133. DOI: 10.1121/1.396393.

[CD17] S. D. Conte et al. *Elementary numerical analysis: an algorithmic approach.* SIAM, 2017.

[Cru+92] C. Cruz-Neira et al. "The CAVE: audio visual experience automatic virtual environment". In: *Communications of the ACM* 35.6 (1992), pp. 64–72.

[Dat10] B. N. Datta. *Numerical linear algebra and applications.* Vol. 116. Siam, 2010.

[De +96] M. De Hoon et al. "Why Yule-Walker should not be used for autoregressive modelling". In: *Annals of nuclear energy* 23.15 (1996), pp. 1219–1228.

[Def+07] J. Defrance et al. "Outdoor sound propagation reference model developed in the European Harmonoise project". In: *Acta Acustica united with Acustica* 93.2 (2007), pp. 213–227.

[Ehr+20] J. Ehret et al. "Evaluating the Influence of Phoneme-Dependent Dynamic Speaker Directivity of Embodied Conversational Agents' Speech". In: *Proceedings of the 20th ACM International Conference on Intelligent Virtual Agents : IVA 2020.* 20th ACM International Conference on Intelligent Virtual Agents (Virtual Event Scotland UK). Association for Computing Machinery, 2020. DOI: 10.1145/3383652.3423863.

[ESV21] A. Erraji et al. "The image edge model". In: *Acta Acustica* 5 (2021), p. 17. DOI: 10.1051/aacus/2021010.

[Far10] A. Farnell. *Designing sound.* Mit Press, 2010.

[FZ07] H. Fastl et al. *Psychoacoustics.* Springer, 2007.

[FBH97] H. Fouad et al. "Perceptually based scheduling algorithms for real-time synthesis of complex sonic environments". In: *Proceedings of the International Conference on Auditory Display.* 1997, pp. 2–5.

[FNH99] H. Fouad et al. "A real-time parallel scheduler for the imprecise computation model". In: *PDCP, Special Issue on Parallel and Distributed Real-Time Systems* 2.1 (1999), pp. 15–23.

[FBB00] H. Fouad et al. "An extensible toolkit for creating virtual sonic environments". In: Georgia Institute of Technology. 2000.

[Fun+04] T. Funkhouser et al. “A beam tracing method for interactive architectural acoustics”. In: *J. Acoust. Soc. Am.* 115 (2004), p. 739.

[GHK19] F. Georgiou et al. “Auralization of a Car Pass-By Using Impulse Responses Computed with a Wave-Based Method”. In: *Acta Acustica united with Acustica* 105.2 (2019), pp. 381–391.

[Ger83] M. A. Gerzon. “Ambisonics in multichannel broadcasting and video”. In: *Audio Engineering Society Convention 74*. Audio Engineering Society. 1983.

[GLH19] G. Grimm et al. “A toolbox for rendering virtual acoustic environments in the context of audiology”. In: *Acta Acustica united with Acustica* 105.3 (2019), pp. 566–578.

[Her99] J. Herder. “Optimization of sound spatialization resource management through clustering”. In: *The Journal of Three Dimensional Images, 3D-Forum Society*. Vol. 13. 3. 1999, pp. 59–65.

[Heu09] K. Heutschi. “Calculation of reflections in an urban environment”. In: *Acta Acustica united with Acustica* 95.4 (2009), pp. 644–652.

[Heu+14] K. Heutschi et al. “Auralization of wind turbine noise: Propagation filtering and vegetation noise synthesis”. In: *ACTA Acustica united with Acustica* 100.1 (2014), pp. 13–24.

[HN06] M. Hodgson et al. “Experimental evaluation of radiosity for room sound-field prediction”. In: *The Journal of the Acoustical Society of America* 120.2 (2006), pp. 808–819.

[Hor09] M. Hornikx. *Numerical modelling of sound propagation to closed urban courtyards*. Chalmers University of Technology Gothenburg, Sweden, 2009.

[Hor16] M. Hornikx. “Ten questions concerning computational urban acoustics”. In: *Building and Environment* 106 (2016), pp. 409–421. DOI: 10.1016/j.buildenv.2016.06.028.

[HP06] S. Hwang et al. “Time delay estimation from HRTFs and HRIRs”. In: *The 8th International Conference on Motion and Vibration Control (MOVIC 2006)*. 2006.

[ISO97] ISO 266:1997. *Acoustics – Preferred frequencies*. International Organization for Standardization (ISO), Geneva, Switzerland. 1997.

[ISO93] ISO 9613-1:1993. *Acoustics – Attenuation of sound during propagation outdoors – Part 1: Calculation of the absorption of sound by the atmosphere*. International Organization for Standardization (ISO), Geneva, Switzerland. 1993.

[ISO96] ISO 9613-2:1996. *Acoustics – Attenuation of sound during propagation outdoors – Part 2: General method of calculation*. International Organization for Standardization (ISO), Geneva, Switzerland. 1996.

[Jek04] U. Jekosch. “Basic Concepts and Terms of”. In: *acta acustica united with Acustica* 90.6 (2004), pp. 999–1006.

[Jol+15] A. Jolibois et al. "Innovative tools for urban soundscape quality: real-time road traffic auralization and low height noise barriers". In: *The Journal of the Acoustical Society of America* 137.4 (2015), pp. 2256–2256.

[Jud+16] T. Judd et al. "Uncertainty-based diffraction using sound particle methods in noise control software". In: *INTER-NOISE and NOISE-CON Congress and Conference Proceedings*. Vol. 253. 5. Institute of Noise Control Engineering. 2016, pp. 3842–3850.

[Kam+18] M. Kamrath et al. "Extending standard urban outdoor noise propagation models to complex geometries". In: *The Journal of the Acoustical Society of America* (2018). DOI: 10.1121/1.5027826.

[Kan02] J. Kang. "Numerical modelling of the sound fields in urban streets with diffusely reflecting boundaries". In: *Journal of Sound and Vibration* 258.5 (2002), pp. 793–813. DOI: 10.1006/jsvi.2002.5150.

[Kan00] J. Kang. "Sound propagation in street canyons: Comparison between diffusely and geometrically reflecting boundaries". In: *The Journal of the Acoustical Society of America* 107.3 (2000), pp. 1394–1404. DOI: 10.1121/1.428580.

[Kan06] J. Kang. *Urban Sound Environment*. Taylor & Francis, 2006.

[KB20] J. M. Kates et al. "Adding air absorption to simulated room acoustic models". In: *The Journal of the Acoustical Society of America* 148.5 (2020), EL408–EL413. DOI: 10.1121/10.0002489.

[Kaw81] T. Kawai. "Sound diffraction by a many-sided barrier or pillar". In: *Journal of Sound Vibration* 79 (1981), pp. 229–242. DOI: 10.1016/0022-460X(81)90370-9.

[Kel57] J. B. Keller. "Diffraction by an Aperture". In: *Journal of Applied Physics* 28.4 (1957), pp. 426–444. DOI: 10.1063/1.1722767.

[Kel62] J. B. Keller. "Geometrical Theory of Diffraction". In: *J. Opt. Soc. Am.* 52.2 (1962), pp. 116–130. DOI: 10.1364/JOSA.52.000116.

[KB51] J. B. Keller et al. "Diffraction and reflection of pulses by wedges and corners". In: *Communications on Pure and Applied Mathematics* 4.1 (1951), pp. 75–94. DOI: 10.1002/cpa.3160040109.

[Kep+14] S. Kephalopoulos et al. "Advances in the development of common noise assessment methods in Europe: The CNOSSOS-EU framework for strategic environmental noise mapping". In: *Science of the Total Environment* 482 (2014), pp. 400–410.

[KC80] W. A. Kinney et al. "A comparison between the Biot-Tolstoy exact solution for wedge diffraction and a related solution that involves the Kirchhoff approximation". In: *J. Acoust. Soc. Am.* 68.S1 (1980), S1–S1. DOI: 10.1121/1.2004619.

[KDS93] M. Kleiner et al. "Auralization - an overview". In: *Journal of the Audio Engineering Society* 41.11 (1993), pp. 861–875.

[Koh+16a] M. Kohnen et al. "Dynamic CTC with and without compensation of early reflections". In: *2016 AES International Conference on Sound Field Control.* International Conference on Sound Field Control, Guildford (UK). 2016.

[Koh+16b] M. Kohnen et al. "Interaktiver Zugang zur Phänomenologie der Akustik in der Hochschullehre". In: *Fortschritte der Akustik : DAGA 2016.* 42. Jahrestagung für Akustik, Aachen (Germany). Deutsche Gesellschaft für Akustik e.V. (DEGA), 2016, pp. 530–533.

[Koh+16c] M. Kohnen et al. "Investigation of dynamic CTC reproduction for rooms with reflecting walls". In: *9th Iberian Congress on Acoustics.* 9th Iberian Congress on Acoustics, Porto (Portugal). 2016.

[KSV17] M. Kohnen et al. "Subjective evaluation of a room-compensated crosstalk cancellation system". In: *Fortschritte der Akustik : DAGA 2017.* 43. Jahrestagung für Akustik, Kiel (Germany). Deutsche Gesellschaft für Akustik e.V. (DEGA), 2017, pp. 1069–1072.

[KP74] R. G. Kouyoumjian et al. "A uniform geometrical theory of diffraction for an edge in a perfectly conducting surface". In: *Proceedings of the IEEE* 62.11 (1974), pp. 1448–1461.

[KSS68] A. Krokstad et al. "Calculating the acoustical room response by the use of a ray tracing technique". In: *Journal of Sound and Vibration* 8.1 (1968), pp. 118–125.

[KA71] U. Kurze et al. "Sound attenuation by barriers". In: *Applied Acoustics* 4.1 (1971), pp. 35–53.

[Laa+96] T. I. Laakso et al. "Splitting the unit delay [FIR/all pass filters design]". In: *IEEE Signal Processing Magazine* 13.1 (1996), pp. 30–60. DOI: 10.1109/79.482137.

[LVK02] P. Larsson et al. "Better Presence and Performance in Virtual Environments by Improved Binaural Sound Rendering". In: *Audio Engineering Society Conference: 22nd International Conference: Virtual, Synthetic, and Entertainment Audio.* 2002.

[LCM07] C. Lauterbach et al. "Interactive sound rendering in complex and dynamic scenes using frustum tracing". In: *Visualization and Computer Graphics, IEEE Transactions on* 13.6 (2007), pp. 1672–1679.

[Leh93] H. Lehnert. "Real-time generation of interactive virtual auditory environments". In: *Proceedings of IEEE Workshop on Applications of Signal Processing to Audio and Acoustics.* IEEE. 1993, pp. 106–109.

[Len06] T. Lentz. "Dynamic crosstalk cancellation for binaural synthesis in virtual reality environments". In: *Journal of the Audio Engineering Society* 54.4 (2006), pp. 283–294.

[Len+07] T. Lentz et al. "Virtual reality system with integrated sound field simulation and reproduction". In: *EURASIP J. Appl. Signal Process.* 2007.1 (2007), pp. 187–187. DOI: 10.1155/2007/70540.

[LHS01] T. Lokki et al. “A framework for evaluating virtual acoustics environments”. English. In: *110th Audio Engineering Society (AES) Convention, Amsterdam the Netherlands, May 12-15 2001*. 2001.

[Lok+02] T. Lokki et al. *Physically-based auralization: design, implementation, and evaluation*. Helsinki University of Technology, 2002.

[LSS02] T. Lokki et al. “An efficient auralization of edge diffraction”. In: *Audio Engineering Society Conference: 21st International Conference: Architectural Acoustics and Sound Reinforcement*. Audio Engineering Society. 2002.

[Mac15] H. M. Macdonald. “A Class of Diffraction Problems”. In: *Proceedings of the London Mathematical Society* s2_14.1 (1915), pp. 410–427. DOI: 10.1112/plms/s2_14.1.410.

[Mac02] H. M. Macdonald. *Electric Waves*. the University Press, 1902.

[Mae68] Z. Maekawa. “Noise reduction by screens”. In: *Applied Acoustics* 1.3 (1968), pp. 157–173. DOI: 10.1016/0003-682X(68)90020-0.

[MD07] D. van Maercke et al. “Development of an analytical model for outdoor sound propagation within the Harmonoise project”. In: *Acta Acustica united with Acustica* 93.2 (2007), pp. 201–212.

[MJ13] J. Maillard et al. “Real time auralization of non-stationary traffic noise-quantitative and perceptual validation in an urban street”. In: *Proc. of AIA-DAGA* (2013).

[Man+12] M. Manyoky et al. “Concept for collaborative design of wind farms facilitated by an interactive GIS-based visual-acoustic 3D simulation”. In: *Proceedings of Digital Landscape Architecture* (2012).

[Mar+18] S. R. Martin et al. “Modeling sound scattering using a combination of the edge source integral equation and the boundary element method”. In: *The Journal of the Acoustical Society of America* 144.1 (2018), pp. 131–141.

[MPM90] D. McNamara et al. “The Uniform Geometrical Theory of Diffraction”. In: *Artech House, London* (1990).

[MV17] J. Mecking et al. “Efficient Simulation of Sound Paths in the Atmosphere”. In: *INTER-NOISE 2017*. 46th International Congress and Exposition on Noise Control Engineering, Hong Kong. Polytechnic University of Hong Kong, 2017, pp. 193–198.

[Med81] H. Medwin. “Shadowing by finite noise barriers”. In: *J. Acoust. Soc. Am.* 69.4 (1981), pp. 1060–1064.

[MCJ82] H. Medwin et al. “Impulse studies of double diffraction: A discrete Huygens interpretation”. In: *J. Acoust. Soc. Am.* 72.3 (1982), pp. 1005–1013. DOI: 10.1121/1.388231.

[Meh+12] R. Mehra et al. “A real-time sound propagation system for noise prediction in outdoor spaces”. In: *INTER-NOISE and NOISE-CON Congress and Conference Proceedings*. Vol. 2012. 4. Citeseer. 2012, pp. 7026–7035.

[Men+18] F. Meng et al. "A concept of auralizing urban pass-by vehicles by measurement- and simulation-based synthesis of sound sources and impulse responses". In: *Proceedings of the 10th European conference on noise control : Euronoise 2018*. 10th European conference on noise control, Crete (Greece). IACM, FORTH, 2018.

[Moe+07] T. Moeck et al. "Progressive Perceptual Audio Rendering of Complex Scenes". Anglais. In: *Symposium on Interactive 3D graphics and games (I3D 2007)*. ACM SIGGRAPH. ACM, 2007, pp. 189–196. DOI: 10.1145/1230100.1230133.

[Møl+95] H. Møller et al. "Head-related transfer functions of human subjects". In: *Journal of the Audio Engineering Society* 43.5 (1995), pp. 300–321.

[Moo83] F. R. Moore. "A general model for spatial processing of sounds". In: *Computer Music Journal* 7.3 (1983), pp. 6–15.

[MVS19] I. Muhammad et al. "Audio-video virtual reality environments in building acoustics: An exemplary study reproducing performance results and subjective ratings of a laboratory listening experiment". In: *The Journal of the Acoustical Society of America* 146.3 (2019), EL310–EL316. DOI: 10.1121/1.5126598.

[NAS08] J. Nam et al. "A method for estimating interaural time difference for binaural synthesis". In: *Audio Engineering Society Convention 125*. Audio Engineering Society. 2008.

[Nay92] G. M. Naylor. "Treatment of early and late reflections in a hybrid computer model for room acoustics". In: *124th ASA Meeting*. 1992.

[OL07] J.-R. Ohm et al. *Signalübertragung - Grundlagen der digitalen und analogen Nachrichtenübertragungssysteme*. Springer, 2007.

[Paa+17] B. Paas et al. "Modelling of Urban Near-Road Atmospheric PM Concentrations Using an Artificial Neural Network Approach with Acoustic Data Input". In: *Environments* 4.2 (2017), p. 26. DOI: 10.3390/environments4020026.

[Paa+18] B. Paas et al. "Modeling of urban near-road atmospheric pollutant concentrations using an Artificial Neural Network approach with novel model input data". In: DMG Mettools X 2018, Braunschweig, Braunschweig (Germany). 2018. DOI: 10.13140/RG.2.2.21980.85120.

[Pau+19] F. Pausch et al. "Implementation and Application of Acoustic Crosstalk Cancellation Systems". In: *Fortschritte der Akustik : DAGA 2019*. 45. Jahrestagung für Akustik, Rostock (Germany). Deutsche Gesellschaft für Akustik e.V. (DEGA), 2019.

[Pel+14] S. Pelzer et al. "Integrating Real-Time Room Acoustics Simulation into a CAD Modeling Software to Enhance the Architectural Design Process". In: *Buildings* 4.2 (2014), pp. 113–138. DOI: 10.3390/buildings4020113.

[PSV] S. Pelzer et al. "The number of necessary rays in geometrically based simulations using the diffuse rain technique". In.

[Pep+11] A. Peplow et al. "Exterior Auralization of Traffic Noise within the LISTEN project". In: *Proceedings of the European Conference on Acoustics (Forum Acusticum 2011), Aalborg, Denmark.* Vol. 27. 2011, pp. 665–669.

[Pie74] A. D. Pierce. "Diffraction of sound around corners and over wide barriers". In: *J. Acoust. Soc. Am.* 55.5 (1974), pp. 941–955. DOI: 10.1121/1.1914668.

[Pie+14] R. Pieren et al. "Auralization of wind turbine noise: Emission synthesis". In: *Acta Acustica united with Acustica* 100.1 (2014), pp. 25–33.

[PBH15] R. Pieren et al. "Auralisation of accelerating passenger cars". In: *Proceedings of the 10th European Conference on Noise Control, Maastricht, The Netherlands.* Vol. 31. 2015, pp. 757–762.

[Pie+16] R. Pieren et al. "Auralisation of Railway Noise: A Concept for the Emission Synthesis of Rolling and Impact Noise". In: *INTER-NOISE and NOISE-CON Congress and Conference Proceedings.* Vol. 253. 8. Institute of Noise Control Engineering. 2016, pp. 274–280.

[PBH16] R. Pieren et al. "Auralization of accelerating passenger cars using spectral modeling synthesis". In: *Applied Sciences* 6.1 (2016), p. 5.

[Pie+19] R. Pieren et al. "Improving future low-noise aircraft technologies using experimental perception-based evaluation of synthetic flyovers". In: *Science of The Total Environment* 692 (2019), pp. 68–81. DOI: 10.1016/j.scitotenv.2019.07.253.

[PK06] B. Plovsing et al. "Nord2000. Comprehensive outdoor sound propagation model. Part 1: Propagation in an atmosphere without significant refraction". In: *DELTA Acoustics & Vibration, Report AV* (2006).

[Pul97] V. Pulkki. "Virtual sound source positioning using vector base amplitude panning". In: *Journal of the audio engineering society* 45.6 (1997), pp. 456–466.

[PLS02] V. Pulkki et al. "Implementation and Visualization of Edge Diffraction with Image-source Method". In: *Audio Engineering Society Convention 112.* 2002.

[Raf04] B. Rafaely. "Plane-wave decomposition of the sound field on a sphere by spherical convolution". In: *The Journal of the Acoustical Society of America* 116.4 (2004), pp. 2149–2157.

[RS14] N. Raghuvanshi et al. "Parametric wave field coding for precomputed sound propagation". In: *ACM Transactions on Graphics (TOG)* 33.4 (2014), pp. 1–11.

[RS18] N. Raghuvanshi et al. "Parametric directional coding for precomputed sound propagation". In: *ACM Transactions on Graphics (TOG)* 37.4 (2018), pp. 1–14.

[Rai98] D. R. Raichel. *Encyclopedia of Acoustics.* 1998.

[Red40] S. Redfearn. "Some acoustical source-observer problems". In: *Philosophical Magazine* 30.200 (1940), pp. 223–236.

[Röc+16] E. Röcher et al. "Dynamic Crosstalk-Cancellation with Room Compensation for Immersive CAVE-Environments". In: *Fortschritte der Akustik : DAGA 2016.* Fortschritte der Akustik. 42. Jahrestagung für Akustik, Aachen (Germany). Deutsche Gesellschaft für Akustik e.V. (DEGA), 2016.

[RSL77] D. J. Rosenkrantz et al. "An analysis of several heuristics for the traveling salesman problem". In: *SIAM journal on computing* 6.3 (1977), pp. 563–581.

[Run+16] A. Rungta et al. "SynCoPation: Interactive Synthesis-Coupled Sound Propagation." In: *IEEE transactions on visualization and computer graphics* (2016).

[Sah+12] A. Sahai et al. "Interdisciplinary Auralization of Take-off and Landing Procedures for Subjective Assessment in Virtual Reality Environments". In: *18th AIAA/CEAS Aeroacoustics Conference (33rd AIAA Aeroacoustics Conference).* 2012, pp. 4–6.

[Sah+16] A. Sahai et al. "Interactive simulation of aircraft noise in aural and visual virtual environments". In: *Applied Acoustics* 101 (2016), pp. 24–38. DOI: 10.1016/j.apacoust.2015.08.002.

[Sal+11] E. Salomons et al. "The Harmonoise sound propagation model". In: *Acta acustica united with acustica* 97.1 (2011), pp. 62–74.

[Sav+99] L. Savioja et al. "Creating interactive virtual acoustic environments". In: *Journal of the Audio Engineering Society* 47.9 (1999), pp. 675–705.

[SS15] L. Savioja et al. "Overview of geometrical room acoustic modeling techniques". In: *The Journal of the Acoustical Society of America* 138.2 (2015), pp. 708–730.

[Sch+17] P. Schäfer et al. "Numerische Stabilität einer Ray Tracing Simulation für atmosphärische Schallfelder". In: *Fortschritte der Akustik : DAGA 2017.* 43. Jahrestagung für Akustik, Kiel (Germany). Deutsche Gesellschaft für Akustik e.V. (DEGA), 2017.

[SMM14] C. Schissler et al. "High-order diffraction and diffuse reflections for interactive sound propagation in large environments". In: *ACM Transactions on Graphics (TOG)* 33.4 (2014), p. 39.

[SM16a] C. Schissler et al. "Adaptive impulse response modeling for interactive sound propagation". In: *Proceedings of the 20th ACM SIGGRAPH Symposium on Interactive 3D Graphics and Games.* ACM. 2016, pp. 71–78.

[SM16b] C. Schissler et al. "Interactive sound propagation and rendering for large multi-source scenes". In: *ACM Transactions on Graphics (TOG)* 36.1 (2016), p. 2.

[Sch+15] T. Schmidt et al. "Psychophysical observations on human perceptions of climatological stress factors in urban environment". In: *Proceedings 19th Triennial Congress of the IEA.* 19th Triennial Congress of the IEA, Melbourne (Australia). 2015, 8 Seiten.

[SP09] D. Schröder et al. "Real-time hybrid simulation method including edge diffraction". In: *EAA auralization symposium* 39 (2009).

[SV07] D. Schröder et al. "Hybrid method for room acoustic simulation in real-time". In: *Proceedings of the 19th International Congress on Acoustics.* 2007.

[Sch+10] D. Schröder et al. "Virtual reality system at RWTH Aachen University". In: *Proceedings of the International Symposium on Room Acoustics (ISRA), Melbourne, Australia.* 2010.

[SSV11] D. Schröder et al. "Open Measurements of Edge Diffraction from a Noise Barrier Scale Model". In: *Building Acoustics* 18.1-2 (2011), pp. 47–57. DOI: 10.1260/1351-010X.18.1-2.47.

[Sch+12] D. Schröder et al. "On the accuracy of edge diffraction simulation methods in geometrical acoustics". In: *Proceedings of the 41st International Congress and Exposition on Noise Control Engineering (Inter-noise), New York, NY, USA.* 2012, pp. 19–22.

[Sil+07] S. Siltanen et al. "The room acoustic rendering equation". In: *The Journal of the Acoustical Society of America* 122.3 (2007), pp. 1624–1635. DOI: 10.1121/1.2766781.

[SSN04] A. Silzle et al. "IKA-SIM: A system to generate auditory virtual environments". In: *Audio Engineering Society Convention 116.* Audio Engineering Society. 2004.

[Smi96] J. O. Smith. "Physical modeling synthesis update". In: *Computer Music Journal* 20.2 (1996), pp. 44–56.

[Smi07] J. O. Smith. *Introduction to Digital Filters with Audio Applications.* W3K Publishing, 2007.

[Smi+97] S. W. Smith et al. *The scientist and engineer's guide to digital signal processing.* Vol. 14. California Technical Pub. San Diego, 1997.

[Som96] A. Sommerfeld. "Mathematische Theorie der Diffraction". German. In: *Mathematische Annalen* 47.2-3 (1896), pp. 317–374. DOI: 10.1007/BF01447273.

[Ste96] U. M. Stephenson. "Quantized pyramidal beam tracing-a new algorithm for room acoustics and noise immission prognosis". In: *Acta Acustica united with Acustica* 82.3 (1996), pp. 517–525.

[Ste10a] U. M. Stephenson. "An analytically derived sound particle diffraction model". In: *Acta Acustica united with Acustica* 96.6 (2010), pp. 1051–1068.

[Ste10b] U. M. Stephenson. "An energetic approach for the simulation of diffraction within ray tracing based on the uncertainty relation". In: *Acta Acustica united with Acustica* 96.3 (2010), pp. 516–535.

[SS07] U. M. Stephenson et al. "An improved energetic approach to diffraction based on the uncertainty principle". In: *19th Int. Cong. on Acoustics (ICA)* (2007).

[Sti+14] J. Stienen et al. "Evaluation of combined stress factors on humans in urban areas". In: *Proceedings of the TECNIACÚSTICA 2014.* 45º Congreso Español de Acústica, Murcia (Spain). 2014, AAM–3 006, 344–350.

[SV15] J. Stienen et al. “Auralization of Urban Environments : concepts towards New Applications”. In: *Proceedings of the 10th European Congress and Exposition on Noise Control Engineering : Euronoise 2015.* 10th European Congress and Exposition on Noise Control Engineering, Maastricht (Netherlands). 2015, pp. 775–780.

[Sti+15a] J. Stienen et al. “Interdisziplinäre Untersuchung zur Wahrnehmung von Lärm und anderen Stressfaktoren im urbanen Raum”. In: *Fortschritte der Akustik : DAGA 2015.* 41. Jahrestagung für Akustik, Nürnberg (Germany). Deutsche Gesellschaft für Akustik e.V. (DEGA), 2015, pp. 1533–1536.

[Sti+15b] J. Stienen et al. “Noise as a Stress Factor on Humans in Urban Environments in Summer and Winter”. In: *Proceedings of the 10th European Congress and Exposition on Noise Control Engineering : Euronoise 2015.* 10th European Congress and Exposition on Noise Control Engineering, Maastricht (Netherlands). 2015, pp. 2445–2450.

[SV16] J. Stienen et al. “Bestimmung von Beugungsparametern für die Echtzeit-Auralisierung in urbanen Räumen”. In: *Fortschritte der Akustik : DAGA 2016.* 42. Jahrestagung für Akustik, Aachen (Germany). Deutsche Gesellschaft für Akustik e.V. (DEGA), 2016.

[SA17] J. Stienen et al. “Akustische Simulation und Wiedergabe für die VR”. In: VR in Industry, Aachen (Germany), 4 Jul 2017 - 5 Jul 2017. 2017.

[SMV17] J. Stienen et al. “Chances and limitations of outdoor sound recordings for interactive Virtual Acoustics”. In: *Fortschritte der Akustik : DAGA 2017.* 43. Jahrestagung für Akustik, Kiel (Germany). Deutsche Gesellschaft für Akustik e.V. (DEGA), 2017.

[SV17] J. Stienen et al. “Geometry-based diffraction auralization for real-time applications in environmental noise”. In: *The Journal of the Acoustical Society of America* 141.5 (2017), pp. 3536–3536. DOI: 10.1121/1.4987468.

[SV18] J. Stienen et al. “Real-time auralization of propagation paths with reflection, diffraction and Doppler shift”. In: *Fortschritte der Akustik : DAGA 2018.* 44. Jahrestagung für Akustik, München (Germany). Deutsche Gesellschaft für Akustik e.V. (DEGA), 2018.

[SV19a] J. Stienen et al. “Progressive region-of-interest filtering for urban sound auralization applications with multiple reflected and diffracted propagation paths”. In: *Proceedings of the 23rd International Congress on Acoustics : ICA 2019.* 23rd International Congress on Acoustics : integrating 4th EAA Euroregio 2019, Aachen (Germany). Deutsche Gesellschaft für Akustik e.V. (DEGA), 2019, pp. 1713–1720. DOI: 10.18154/RWTH-CONV-239887.

[SAV19] J. Stienen et al. “Strategies for the auralization of simulated urban traffic”. In: *Proceedings of Internoise 2019 : Internoise 2019.* Vol. 259. 9. 48th International Congress and Exhibition onNoise Control Engineering, Madrid (Spain). 2019, pp. 10–10.

[SV19b] J. Stienen et al. "The open-source Virtual Acoustics (VA) real-time auralization framework". In: *Proceedings of the 23rd International Congress on Acoustics : ICA 2019.* 23rd International Congress on Acoustics : integrating 4th EAA Euroregio 2019, Aachen (Germany). Deutsche Gesellschaft für Akustik e.V. (DEGA), 2019.

[Str98] H. Strauss. "Implementing Doppler Shifts for Virtual Auditory Environments". In: *Audio Engineering Society Convention 104, Amsterdam, NL.* 1998.

[SFV99] P. Svensson et al. "An analytic secondary source model of edge diffraction impulse responses". In: *J. Acoust. Soc. Am.* 106 (1999), p. 2331.

[SC06] U. P. Svensson et al. "Edge-Diffraction Impulse Responses Near Specular-Zone and Shadow-Zone Boundaries". In: *Acta Acustica united with Acustica* 92.4 (2006), pp. 501–512.

[SA12] U. P. Svensson et al. "An edge-source integral equation for the calculation of scattering". In: *J. Acoust. Soc. Am.* 132.3 (2012), pp. 1889–1889. DOI: 10.1121/1.4754928.

[TH92] T. Takala et al. "Sound rendering". In: *Proceedings of the 19th annual conference on Computer graphics and interactive techniques.* 1992, pp. 211–220.

[Tay+09a] M. Taylor et al. "Fast edge-diffraction for sound propagation in complex virtual environments". In: *EAA auralization symposium.* 2009, pp. 15–17.

[Tay+12] M. Taylor et al. "Guided multiview ray tracing for fast auralization". In: *IEEE Transactions on Visualization and Computer Graphics* 18 (2012), pp. 1797–1810.

[Tay+09b] M. T. Taylor et al. "Resound: interactive sound rendering for dynamic virtual environments". In: *Proceedings of the 17th ACM international conference on Multimedia.* 2009, pp. 271–280.

[Tho+12] P. Thomas et al. "Auralisation of a car pass-by behind a low finite-length vegetated noise barrier". In: *Proceedings of the 9th European Conference on Noise Control.* 2012, pp. 932–937.

[TRK04] R. Torres et al. "Edge diffraction and surface scattering in concert halls: physical and perceptual aspects". In: *Journal of Temporal Design in Architecture and the Environment* 4.1 (2004), pp. 52–58.

[TK98] R. R. Torres et al. "Audibility of Edge Diffraction in Auralization of a Stage House". In: *Proceedings of the 16th International Congress on Acoustics.* 1998.

[TSK01] R. R. Torres et al. "Computation of edge diffraction for more accurate room acoustics auralization". In: *J. Acoust. Soc. Am.* 109.2 (2001), pp. 600–610. DOI: 10.1121/1.1340647.

[TF07] C. Tsakostas et al. "Real-time spatial representation of moving sound sources". In: *Audio Engineering Society Convention 123.* Audio Engineering Society. 2007.

[Tsi07] N. Tsingos. "Perceptually-based auralization". In: *19th International Congress on Acoustics*. 2007.

[TG97] N. Tsingos et al. "Soundtracks for Computer Animation : Sound Rendering in Dynamic Environments with Occlusions". English. In: *Graphics Interface '97*. 1997.

[Tsi+01] N. Tsingos et al. "Modeling Acoustics in Virtual Environments Using the Uniform Theory of Diffraction". In: *Proceedings of the 28th Annual Conference on Computer Graphics and Interactive Techniques*. SIGGRAPH '01. ACM, 2001, pp. 545–552. DOI: 10.1145/383259.383323.

[TGD04] N. Tsingos et al. "Perceptual audio rendering of complex virtual environments". In: *ACM Transactions on Graphics (TOG)*. Vol. 23. ACM. 2004, pp. 249–258.

[VSB06] T. Van Renterghem et al. "Parameter study of sound propagation between city canyons with a coupled FDTD-PE model". In: *Applied Acoustics* 67.6 (2006), pp. 487–510.

[Van+12] T. Van Renterghem et al. "Road traffic noise reduction by vegetated low noise barriers in urban streets". In: *Proceedings of the 9th European conference on noise control : Euronoise 2012*. 9th European conference on noise control, Prague. 2012.

[Vat09] A. Vattani. "The hardness of k-means clustering in the plane". In: *Manuscript, accessible at http://cseweb.ucsd.edu/avattani/papers/kmeans_hardness.pdf* 617 (2009).

[VVO+15] E. M. Viggen et al. "Development of an outdoor auralisation prototype with 3D sound reproduction". In: *18th International Conference on Digital Audio Effects (DAFx-15)*. 2015.

[Vor11] M. Vorländer. *Auralization: Fundamentals of Acoustics, Modelling, Simulation, Algorithms and Acoustic Virtual Reality*. RWTHedition. Springer Berlin Heidelberg, 2011.

[VS15] M. Vorländer et al. "Virtual acoustic environments for soundscape research and urban planning". In: *The journal of the Acoustical Society of America : JASA-O* 138.3 (2015), p. 1748. DOI: 10.1121/1.4933518.

[VAS19] M. Vorländer et al. "An efficient DSP network for the real-time auralization of complex urban scenarios". In: *The Journal of the Acoustical Society of America : JASA-O* 146.4 (2019), p. 2795. DOI: 10.1121/1.5136683.

[WV14] F. Wefers et al. "Zeitvariante Beschreibung virtueller Szenen für die Echtzeit-Auralisierung instationärer Schallfelder". In: *Jahrestagung Deutsche Gesellschaft für Akustik (DAGA 2014), Oldenburg, Germany*. 2014.

[Wef15] F. Wefers. *Partitioned convolution algorithms for real-time auralization*. Vol. 20. Logos Verlag Berlin GmbH, 2015.

[Wef17] F. Wefers. "Berechnung der Schallausbreitungsdauer für beliebige Bewegungsbahnkurven mittels numerischer Lösungsverfahren". In: *Fortschritte der Akustik, DAGA 2017, Kiel* (2017).

[Wef+14] F. Wefers et al. “Interactive Acoustic Virtual Environments using Distributed Room Acoustics Simulation”. In: *Proc. of the EAA Joint Symposium on Auralization and Ambisonics, Berlin, Germany.* 2014. DOI: `10.14279/depositonce-9`.

[WV18] F. Wefers et al. “Flexible data structures for dynamic virtual auditory scenes”. In: *Virtual Reality* (2018), pp. 1–15.

[WAS17] S. Weigand et al. “Evaluation of higher order sound particle diffraction with measurements around finite sized screens in a semi-anechoic chamber”. In: *The Journal of the Acoustical Society of America* 141.5 (2017), pp. 3779–3779. DOI: `10.1121/1.4988313`.

[WSS18] S. Weigand et al. “Simulation of multiple Sound Particle Diffraction based on the Uncertainty Relation-a revolution in noise immission prognosis; Part II: Evaluation by Measurements”. In: *Proceedings of the 10th European conference on noise control : Euronoise 2018.* 10th European conference on noise control, Crete (Greek). 2018.

[Wen+18] J. Wendt et al. “Does the Directivity of a Virtual Agent's Speech Influence the Perceived Social Presence?” In: *IEEE Virtual Humans and Crowds for Immersive Environments : VHCIE 2018.* IEEE Virtual Humans and Crowds for Immersive Environments, Reutlingen (Germany). 2018.

[Wen+19] J. Wendt et al. “Influence of Directivity on the Perception of Embodied Conversational Agents' Speech”. In: *Proceedings of the 19th ACM International Conference on Intelligent Virtual Agents : IVA 2019.* 19th ACM International Conference on Intelligent Virtual Agents, Paris (France). ACM Press New York, 2019, pp. 130–132. DOI: `10.1145/3308532.3329434`.

[Wen98] E. M. Wenzel. “The impact of system latency on dynamic performance in virtual acoustic environments”. In: *Target* 135.180 (1998).

[WMA00] E. M. Wenzel et al. “Sound Lab: A real-time, software-based system for the study of spatial hearing”. In: *Audio Engineering Society Convention 108.* Audio Engineering Society. 2000.

[Wor99] World Health Organization. *Guidelines for Community Noise.* World Health Organization, Regional Office for Europe, 1999.

[Wor11] World Health Organization. *Burden of disease from environmental noise: Quantification of healthy life years lost in Europe.* World Health Organization. Regional Office for Europe, 2011.

[Yeh+13] H. Yeh et al. “Wave-ray coupling for interactive sound propagation in large complex scenes”. In: *ACM Transactions on Graphics (TOG)* 32.6 (2013), p. 165.

[ZM14] H. Ziegelwanger et al. “Modeling the direction-continuous time-of-arrival in head-related transfer functions”. In: *The Journal of the Acoustical Society of America* 135.3 (2014), pp. 1278–1293.

Curriculum Vitae

Personal Data

	Jonas Stienen
14.09.1984	born in Münster, Germany

Education

1989–1991	Primary School Bethel, Lesotho, Africa
1991–1993	Boland-Gemeinschafts-Grundschule, Herzebrock, Germany
1993–1994	Kattenstrother Grundschule, Gütersloh, Germany
1994–2004	Anne-Frank-Gesamtschule, Gütersloh, Germany

Higher Education

2005–2013	Computer Engineering (Diplom) at RWTH Aachen University, Aachen, Germany

Professional Experience

2012	Internship at SVT Engineering Consultants, Perth, Australia
2013–2020	Research Assistant at the Institute of Technical Acoustics (ITA), RWTH Aachen University, Aachen, Germany
2021	HEAD acoustics GmbH, Herzogenrath, Aachen, Germany

Aachen, Germany, June 01, 2021.

Publications

Articles

[ESV21] A. Erraji et al. "The image edge model". In: *Acta Acustica* 5 (2021), p. 17. DOI: 10.1051/aacus/2021010.

[Paa+17] B. Paas et al. "Modelling of Urban Near-Road Atmospheric PM Concentrations Using an Artificial Neural Network Approach with Acoustic Data Input". In: *Environments* 4.2 (2017), p. 26. DOI: 10.3390/environments4020026.

Conferences

[Asp+18] L. Aspöck et al. "Application of virtual acoustic environments in the scope of auditory research". In: *XXVIII Encontro da Sociedade Brasileira de Acústica : SOBRAC 2018.* XXVIII Encontro da Sociedade Brasileira de Acústica, Porto Alegre (Brazil). 2018. DOI: 10.17648/sobrac-87162.

[Asp+19] L. Aspöck et al. "Auralization of interactive virtual scenes containing numerous sound sources". In: *Proceedings of the 23rd International Congress on Acoustics : ICA 2019.* 23rd International Congress on Acoustics : integrating 4th EAA Euroregio 2019, Aachen (Germany). Deutsche Gesellschaft für Akustik e.V. (DEGA), 2019, pp. 6401–6401. DOI: 10.18154/RWTH-CONV-239281.

[Ehr+20] J. Ehret et al. "Evaluating the Influence of Phoneme-Dependent Dynamic Speaker Directivity of Embodied Conversational Agents' Speech". In: *Proceedings of the 20th ACM International Conference on Intelligent Virtual Agents : IVA 2020.* 20th ACM International Conference on Intelligent Virtual Agents (Virtual Event Scotland UK). Association for Computing Machinery, 2020. DOI: 10.1145/3383652.3423863.

[Koh+16b] M. Kohnen et al. "Interaktiver Zugang zur Phänomenologie der Akustik in der Hochschullehre". In: *Fortschritte der Akustik : DAGA 2016.* 42. Jahrestagung für Akustik, Aachen (Germany). Deutsche Gesellschaft für Akustik e.V. (DEGA), 2016, pp. 530–533.

[Koh+16c] M. Kohnen et al. "Investigation of dynamic CTC reproduction for rooms with reflecting walls". In: *9th Iberian Congress on Acoustics.* 9th Iberian Congress on Acoustics, Porto (Portugal). 2016.

[KSV17] M. Kohnen et al. “Subjective evaluation of a room-compensated crosstalk cancellation system”. In: *Fortschritte der Akustik : DAGA 2017.* 43. Jahrestagung für Akustik, Kiel (Germany). Deutsche Gesellschaft für Akustik e.V. (DEGA), 2017, pp. 1069–1072.

[Röc+16] E. Röcher et al. “Dynamic Crosstalk-Cancellation with Room Compensation for Immersive CAVE-Environments”. In: *Fortschritte der Akustik : DAGA 2016.* Fortschritte der Akustik. 42. Jahrestagung für Akustik, Aachen (Germany). Deutsche Gesellschaft für Akustik e.V. (DEGA), 2016.

[Sch+17] P. Schäfer et al. “Numerische Stabilität einer Ray Tracing Simulation für atmosphärische Schallfelder”. In: *Fortschritte der Akustik : DAGA 2017.* 43. Jahrestagung für Akustik, Kiel (Germany). Deutsche Gesellschaft für Akustik e.V. (DEGA), 2017.

[Sch+15] T. Schmidt et al. “Psychophysical observations on human perceptions of climatological stress factors in urban environment”. In: *Proceedings 19th Triennial Congress of the IEA.* 19th Triennial Congress of the IEA, Melbourne (Australia). 2015, 8 Seiten.

[Sti+14] J. Stienen et al. “Evaluation of combined stress factors on humans in urban areas”. In: *Proceedings of the TECNIACÚSTICA 2014.* 45º Congreso Español de Acústica, Murcia (Spain). 2014, AAM-3 006, 344–350.

[SV15] J. Stienen et al. “Auralization of Urban Environments : concepts towards New Applications”. In: *Proceedings of the 10th European Congress and Exposition on Noise Control Engineering : Euronoise 2015.* 10th European Congress and Exposition on Noise Control Engineering, Maastricht (Netherlands). 2015, pp. 775–780.

[Sti+15a] J. Stienen et al. “Interdisziplinäre Untersuchung zur Wahrnehmung von Lärm und anderen Stressfaktoren im urbanen Raum”. In: *Fortschritte der Akustik : DAGA 2015.* 41. Jahrestagung für Akustik, Nürnberg (Germany). Deutsche Gesellschaft für Akustik e.V. (DEGA), 2015, pp. 1533–1536.

[Sti+15b] J. Stienen et al. “Noise as a Stress Factor on Humans in Urban Environments in Summer and Winter”. In: *Proceedings of the 10th European Congress and Exposition on Noise Control Engineering : Euronoise 2015.* 10th European Congress and Exposition on Noise Control Engineering, Maastricht (Netherlands). 2015, pp. 2445–2450.

[SV16] J. Stienen et al. “Bestimmung von Beugungsparametern für die Echtzeit-Auralisierung in urbanen Räumen”. In: *Fortschritte der Akustik : DAGA 2016.* 42. Jahrestagung für Akustik, Aachen (Germany). Deutsche Gesellschaft für Akustik e.V. (DEGA), 2016.

[SV19a] J. Stienen et al. “Progressive region-of-interest filtering for urban sound auralization applications with multiple reflected and diffracted propagation paths”. In: *Proceedings of the 23rd International Congress on Acoustics : ICA 2019.* 23rd International Congress on Acoustics : integrating 4th EAA Euroregio 2019, Aachen (Germany). Deutsche Gesellschaft für Akustik e.V. (DEGA), 2019, pp. 1713–1720. DOI: 10.18154/RWTH-CONV-239887.

[SAV19] J. Stienen et al. “Strategies for the auralization of simulated urban traffic”. In: *Proceedings of Internoise 2019 : Internoise 2019*. Vol. 259. 9. 48th International Congress and Exhibition onNoise Control Engineering, Madrid (Spain). 2019, pp. 10–10.

[Wen+18] J. Wendt et al. “Does the Directivity of a Virtual Agent's Speech Influence the Perceived Social Presence?” In: *IEEE Virtual Humans and Crowds for Immersive Environments : VHCIE 2018*. IEEE Virtual Humans and Crowds for Immersive Environments, Reutlingen (Germany). 2018.

[Wen+19] J. Wendt et al. “Influence of Directivity on the Perception of Embodied Conversational Agents' Speech”. In: *Proceedings of the 19th ACM International Conference on Intelligent Virtual Agents : IVA 2019*. 19th ACM International Conference on Intelligent Virtual Agents, Paris (France). ACM Press New York, 2019, pp. 130–132. DOI: 10.1145/3308532.3329434.

Posters

[Paa+18] B. Paas et al. “Modeling of urban near-road atmospheric pollutant concentrations using an Artificial Neural Network approach with novel model input data”. In: DMG Mettools X 2018, Braunschweig, Braunschweig (Germany). 2018. DOI: 10.13140/RG.2.2.21980.85120.

[SV18] J. Stienen et al. “Real-time auralization of propagation paths with reflection, diffraction and Doppler shift”. In: *Fortschritte der Akustik : DAGA 2018*. 44. Jahrestagung für Akustik, München (Germany). Deutsche Gesellschaft für Akustik e.V. (DEGA), 2018.

[SV19b] J. Stienen et al. “The open-source Virtual Acoustics (VA) real-time auralization framework”. In: *Proceedings of the 23rd International Congress on Acoustics : ICA 2019*. 23rd International Congress on Acoustics : integrating 4th EAA Euroregio 2019, Aachen (Germany). Deutsche Gesellschaft für Akustik e.V. (DEGA), 2019.

Presentations

[Koh+16a] M. Kohnen et al. “Dynamic CTC with and without compensation of early reflections”. In: *2016 AES International Conference on Sound Field Control*. International Conference on Sound Field Control, Guildford (UK). 2016.

[MV17] J. Mecking et al. “Efficient Simulation of Sound Paths in the Atmosphere”. In: *INTER-NOISE 2017*. 46th International Congress and Exposition on Noise Control Engineering, Hong Kong. Polytechnic University of Hong Kong, 2017, pp. 193–198.

[Men+18] F. Meng et al. "A concept of auralizing urban pass-by vehicles by measurement- and simulation-based synthesis of sound sources and impulse responses". In: *Proceedings of the 10th European conference on noise control : Euronoise 2018*. 10th European conference on noise control, Crete (Greece). IACM, FORTH, 2018.

[Pau+19] F. Pausch et al. "Implementation and Application of Acoustic Crosstalk Cancellation Systems". In: *Fortschritte der Akustik : DAGA 2019*. 45. Jahrestagung für Akustik, Rostock (Germany). Deutsche Gesellschaft für Akustik e.V. (DEGA), 2019.

[SA17] J. Stienen et al. "Akustische Simulation und Wiedergabe für die VR". In: VR in Industry, Aachen (Germany), 4 Jul 2017 - 5 Jul 2017. 2017.

[SMV17] J. Stienen et al. "Chances and limitations of outdoor sound recordings for interactive Virtual Acoustics". In: *Fortschritte der Akustik : DAGA 2017*. 43. Jahrestagung für Akustik, Kiel (Germany). Deutsche Gesellschaft für Akustik e.V. (DEGA), 2017.

[SV17] J. Stienen et al. "Geometry-based diffraction auralization for real-time applications in environmental noise". In: *The Journal of the Acoustical Society of America* 141.5 (2017), pp. 3536–3536. DOI: 10.1121/1.4987468.

[VS15] M. Vorländer et al. "Virtual acoustic environments for soundscape research and urban planning". In: *The journal of the Acoustical Society of America : JASA-O* 138.3 (2015), p. 1748. DOI: 10.1121/1.4933518.

[VAS19] M. Vorländer et al. "An efficient DSP network for the real-time auralization of complex urban scenarios". In: *The Journal of the Acoustical Society of America : JASA-O* 146.4 (2019), p. 2795. DOI: 10.1121/1.5136683.

Acknowledgements

I would like to thank Frank Wefers for kindling my enthusiasm for virtual acoustics and Dominik Rausch, who did the same for VR. Your guidance and coding support were tremendously valuable to me and deserve my deepest appreciation.

Without the accomplishments of all the students I supervised, the realisation of an outdoor auralisation module would not have been possible. I can say with great pleasure that the Bachelor theses of Daniel Filbert, Julian Jansen, Amed Selvi as well as the Master theses of Lukas Mösch, Armin Erraji, Henry Andrew, Lennart Reich and Pascal Palenda have delivered a substantial contribution to VA in general and the outdoor auralisation module in particular. A special thanks goes to Armin, as your implementation skills and the energy you have invested into the publication of the IEM were outstanding.

Furthermore, I want to accentuate that I sincerely enjoyed the fruitful cooperation with my colleagues in the interdisciplinary projects I was entrusted with. I am contented that the collaboration is manifested in the list of publications and I want to particularly thank Jonathan Ehret and Bastian Paas for the huge efforts that went into the manuscripts and the review processes of our papers.

Regarding my dissertation, I appreciate the kind support of Johannes Klein and Florian Pausch in technical questions and motivational matters, Lukas Aspöck for the guidance concerning the comparison of the propagation simulation with the BRAS measurements and Stefan Weigand for lively discussions on diffraction issues. All the more, I express my gratitude to the proof reading by Michael Kohnen, Mark Müller-Giebeler and again Lukas Aspöck and Florian Pausch.

Leaving the institute after almost a decade of involvement closes an important chapter of my life. It comes without saying that I will miss the fascinating discussions on acoustics and other topics during coffee breaks and I am thankful for all the small aspects that sharpened the senses to improve our real-time auralisation framework. In this context, especially Philipp Schäfer receives my best regards, as the VA project will continue under his competent supervision.

Additionally, I want to thank the staff of the mechanical and electrical workshops for their invaluable contributions to our VR setups and the secretary's office for being willing to listen to my concerns and making problems disappear.

Florian Pausch, my office colleague and dear friend, it is outside my imagination how the time at the institute might have been without you, but I am most grateful for it.

Last but not least, I want to thank Janina Fels for opening the door for my time at the institute, Torsten Kuhlen for the trust to operate and extend the *aixCAVE* system, Gottfried Behler for countless technical and not so technical discussions as well as Julio Torres and associates for the wonderful collaboration and the most pleasent visits in Brasil.

And finally, I express my deepest gratitude to you, Michael Vorländer, for giving me the opportunity to obtain a doctorate and providing scientific guidance. I hope, that I was able to repay the trust. I immensely admire the incredible responsibilities you assumed for the acoustic community and I have yet to figure out how you make it look so easy.

Bisher erschienene Bände der Reihe

Aachener Beiträge zur Akustik

ISSN 1866-3052
ISSN 2512-6008 (seit Band 28)

1	Malte Kob	Physical Modeling of the Singing Voice ISBN 978-3-89722-997-6 40.50 EUR
2	Martin Klemenz	Die Geräuschqualität bei der Anfahrt elektrischer Schienenfahrzeuge ISBN 978-3-8325-0852-4 40.50 EUR
3	Rainer Thaden	Auralisation in Building Acoustics ISBN 978-3-8325-0895-1 40.50 EUR
4	Michael Makarski	Tools for the Professional Development of Horn Loudspeakers ISBN 978-3-8325-1280-4 40.50 EUR
5	Janina Fels	From Children to Adults: How Binaural Cues and Ear Canal Impedances Grow ISBN 978-3-8325-1855-4 40.50 EUR
6	Tobias Lentz	Binaural Technology for Virtual Reality ISBN 978-3-8325-1935-3 40.50 EUR
7	Christoph Kling	Investigations into damping in building acoustics by use of downscaled models ISBN 978-3-8325-1985-8 37.00 EUR
8	Joao Henrique Diniz Guimaraes	Modelling the dynamic interactions of rolling bearings ISBN 978-3-8325-2010-6 36.50 EUR
9	Andreas Franck	Finite-Elemente-Methoden, Lösungsalgorithmen und Werkzeuge für die akustische Simulationstechnik ISBN 978-3-8325-2313-8 35.50 EUR

10	Sebastian Fingerhuth	Tonalness and consonance of technical sounds ISBN 978-3-8325-2536-1 42.00 EUR
11	Dirk Schröder	Physically Based Real-Time Auralization of Interactive Virtual Environments ISBN 978-3-8325-2458-6 35.00 EUR
12	Marc Aretz	Combined Wave And Ray Based Room Acoustic Simulations Of Small Rooms ISBN 978-3-8325-3242-0 37.00 EUR
13	Bruno Sanches Masiero	Individualized Binaural Technology. Measurement, Equalization and Subjective Evaluation ISBN 978-3-8325-3274-1 36.50 EUR
14	Roman Scharrer	Acoustic Field Analysis in Small Microphone Arrays ISBN 978-3-8325-3453-0 35.00 EUR
15	Matthias Lievens	Structure-borne Sound Sources in Buildings ISBN 978-3-8325-3464-6 33.00 EUR
16	Pascal Dietrich	Uncertainties in Acoustical Transfer Functions. Modeling, Measurement and Derivation of Parameters for Airborne and Structure-borne Sound ISBN 978-3-8325-3551-3 37.00 EUR
17	Elena Shabalina	The Propagation of Low Frequency Sound through an Audience ISBN 978-3-8325-3608-4 37.50 EUR
18	Xun Wang	Model Based Signal Enhancement for Impulse Response Measurement ISBN 978-3-8325-3630-5 34.50 EUR
19	Stefan Feistel	Modeling the Radiation of Modern Sound Reinforcement Systems in High Resolution ISBN 978-3-8325-3710-4 37.00 EUR
20	Frank Wefers	Partitioned convolution algorithms for real-time auralization ISBN 978-3-8325-3943-6 44.50 EUR

21 Renzo Vitale
Perceptual Aspects Of Sound Scattering In Concert Halls
ISBN 978-3-8325-3992-4 34.50 EUR

22 Martin Pollow
Directivity Patterns for Room Acoustical Measurements and Simulations
ISBN 978-3-8325-4090-6 41.00 EUR

23 Markus Müller-Trapet
Measurement of Surface Reflection Properties. Concepts and Uncertainties
ISBN 978-3-8325-4120-0 41.00 EUR

24 Martin Guski
Influences of external error sources on measurements of room acoustic parameters
ISBN 978-3-8325-4146-0 46.00 EUR

25 Clemens Nau
Beamforming in modalen Schallfeldern von Fahrzeuginnenräumen
ISBN 978-3-8325-4370-9 47.50 EUR

26 Samira Mohamady
Uncertainties of Transfer Path Analysis and Sound Design for Electrical Drives
ISBN 978-3-8325-4431-7 45.00 EUR

27 Bernd Philippen
Transfer path analysis based on in-situ measurements for automotive applications
ISBN 978-3-8325-4435-5 50.50 EUR

28 Ramona Bomhardt
Anthropometric Individualization of Head-Related Transfer Functions Analysis and Modeling
ISBN 978-3-8325-4543-7 35.00 EUR

29 Fanyu Meng
Modeling of Moving Sound Sources Based on Array Measurements
ISBN 978-3-8325-4759-2 44.00 EUR

30 Jan-Gerrit Richter
Fast Measurement of Individual Head-Related Transfer Functions
ISBN 978-3-8325-4906-0 45.50 EUR

31 Rhoddy A. Viveros Muñoz
Speech perception in complex acoustic environments: Evaluating moving maskers using virtual acoustics
ISBN 978-3-8325-4963-3 35.50 EUR

32 Spyros Brezas
Investigation on the dissemination of unit watt in airborne sound and applications
ISBN 978-3-8325-4971-8 47.50 EUR

33	Josefa Oberem	Examining auditory selective attention: From dichotic towards realistic environments ISBN 978-3-8325-5101-8 43.00 EUR
34	Johannes Klein	Directional Room Impulse Response Measurement ISBN 978-3-8325-5139-1 41.00 EUR
35	Lukas Aspöck	Validation of room acoustic simulation models ISBN 978-3-8325-5204-6 57.00 EUR
36	Florian Pausch	Spatial audio reproduction for hearing aid research: System design, evaluation and application ISBN 978-3-8325-5461-3 49.50 EUR
37	Ingo Witew	Measurements in room acoustics. Uncertainties and influence of the measurement position ISBN 978-3-8325-5529-0 63.00 EUR
38	Imran Muhammad	Virtual Building Acoustics: Auralization with Contextual and Interactive Features ISBN 978-3-8325-5601-3 55.50 EUR
39	Jonas Stienen	Real-Time Auralisation of Outdoor Sound Propagation ISBN 978-3-8325-5629-7 40.00 EUR

Alle erschienenen Bücher können unter der angegebenen ISBN-Nummer direkt online (http://www.logos-verlag.de) oder per Fax (030 - 42 85 10 92) beim Logos Verlag Berlin bestellt werden.